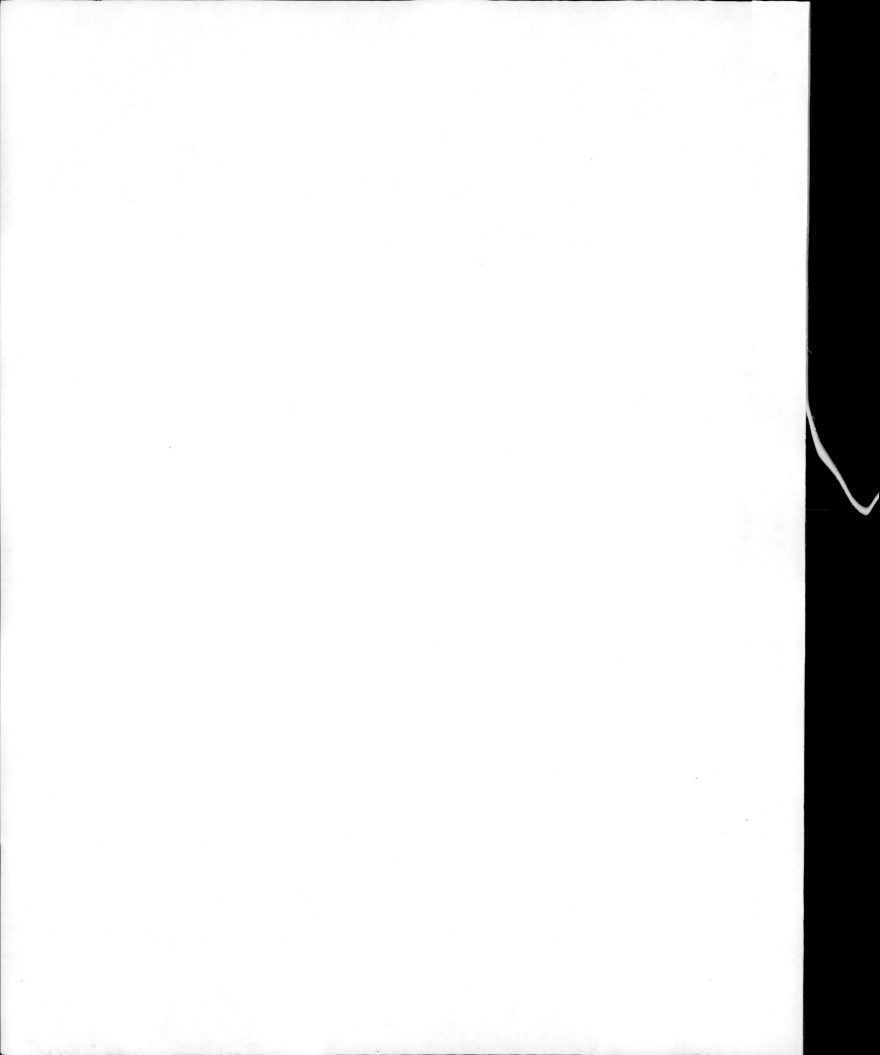

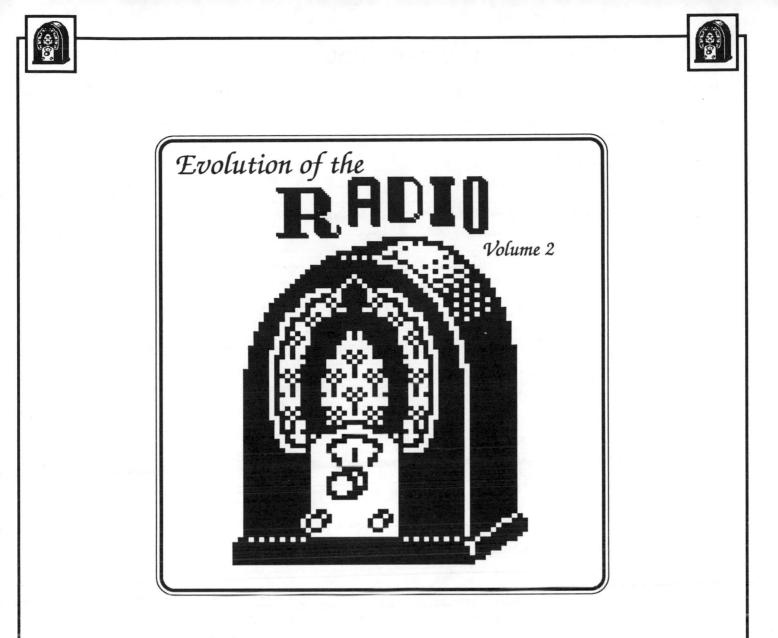

Evolution of the

RADIO

Volume 2

ISBN# 0-89538-019-6

Edited by Scott Wood

L-W Book Sales
P. O. Box 69
Gas City, IN 46933

Layout: David Devon Dilley

Table of Contents

We had no idea when we published Vol. 1 that the response would be so great. Radio collectors around the world have appreciated the format of the first volume and expressed their desire to see more radios in "living color photos". Therefore, here is Vol.2.

We would like to thank all of the fine collectors and dealers who contributed to this book. If you have more photos and info that you would like to send us, we guess there might possibly be a third volume. We'll gladly give you credit in the book as before!!

Format and Pricing Information

In using this book you must remember that information and photos came from many different people from throughout the country. In some cases information may have been a little sketchy and some of the photos low quality. We have tried our very best to include all photos and appreciate all who donated. Most photos are self explanatory so we haven't included information that we felt would be unnecessary. Prices stated in the Price Guide are for radios in fine condition (Not necessarily the radio in the photo). Pricing was based on many factors.... Auction prices, Show prices, Owner's opinions, etc. The prices for the catalog pages and magazine ads are for the radios pictured in them. Remember, this is only a guide and should be used as such. L-W Book Sales is not responsible for gains or losses in using this book.

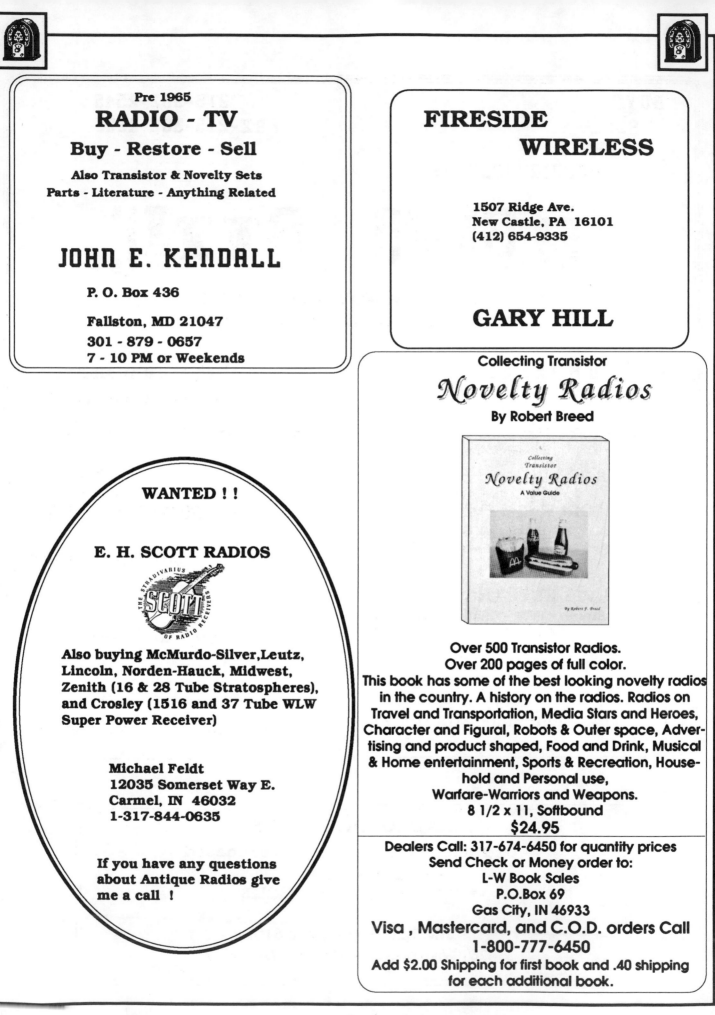

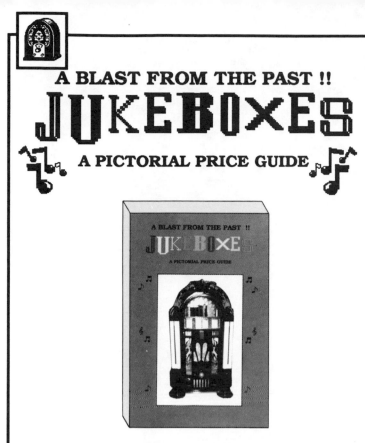
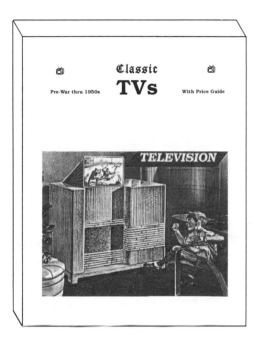

Contributors

Allport, John C., 210 Cleveland, Cary, IL, 60013

Amos, Bob, 35682 Charles St., Wildomar, CA, 92595

Anderson, Larry & Nancy, 176 Oakmont Hills Road, Wheeling, WV 26003

Bart, David, 5324 West Foster, Chicago, IL, 60630

Beaudet, James A., 9908 Woodland NE, Albuquerque, NM ,87112-2211

Berg, Jim, 4261 Wilcox Rd. - Box A, Northport, WA, 99157

Boucher, Ron, P.O. Box 541, Goffstown, NH, 03045

Burskey, Doug, 732 Pulver List Rd., Mansfield, OH, 44905

Byrd, Mark W., 6710 Sutter Park, Houston, TX, 77066

Davis, Richard M., 2655 W. Park Dr., Baltimore, MD, 21207-6040

Deecken, Ken, 1247 S.W. 13th St., Boca Raton, FL, 33486

Doggett, Spencer J., 61333 North Ridge Trail, Washington, MI, 48094

Embry, John H., 1304 Independence Drive, Slidell, LA, 70458

Farmer, Greg, 71 Rice Creek Way, Fridley, MN, 55432

Feldt, Michael, 12035 Somerset Way E., Carmel, IN 46033

Ginocchio, Tom, 719 Carroll St., Brooklyn, NY, 11215

Grossman, Ira, Silver Springs, MD, (301)-460-3981

Hill, Gary, 1507 Ridge Ave., New Castle, PA, 16101

Kaplan, M.A., 14902 84th Ave. Ct. N.W., Gig Harbor, WA, 98329

Kendall, David, 401 Himes St., Huntington, IN, 46750

Kendall, John E., P.O. Box 436, Fallson, MD, 21047

Kinnard, Jay, 623 Amesbury Lane, Austin, TX, 78752

Koste, Mike, 57 Tennis Ave., Ambler, PA, 19002

Kreitner, Arthur, 2013 E. Belle, Belleville, IL, 62221

Kushnir, Ray, 2213 Alpine Drive, Colorado Springs, CO, 80909

LaFleur, Ted, 52333 Ave. Bermudas, La Quinta, CA, 92253

Larsen, Gerald, 7841 W. Elm Grove Dr., Elmwood Park, IL ,60635

Leeth, Carol, 801 S. Webster #2, Anaheim, CA, 92804

Lyons, David C., 623 Carlson St., Sycamore, IL 60178 - 2401

Mac's Old Time Radios, 4335 W. 147th St., Lawndale, CA, 90260

Maldonado, Max, Rt.2 - Box 561, Dallas, NC, 28034

Mason, Ross, 715 S. Penn, Mason City, IA, 50401-5153

Mednick, David, 1450 Palisade Ave. #5-H, Fort Lee, NJ, 07024

Melvin, Steve, 34 Deep Brook Harbor Rd., Suffield, CT, 06078

Nelson, Greg, 1022-9 Wood Creek Dr., Fayetteville, NC, 28314-1120

Nordboe, Don, 3220 W. Broadway, Council Bluffs, IA, 51501

O'Neill, James, 170 Devon Rd., Delaware, OH, 43015-2821

Oppenheim, Peter, 146 E. 49th St. - Apt. 3A, New York, NY, 10017

Osborne, Dennis, P.O. Box 5096, Raleigh, NC, 27650

Peterson, Steve, 5912 W. County Rd. 875 S., Knightstown, IN, 46148

Reinicke, John, 7458 St. Auburn Dr., Bloomfield, MI, 48301

Ricci, Joseph A., 5433 Oak Tree Drive, Arnold, MO, 63010

Richard's Radios of Omaha, P.O. Box 12173, Omaha, NE, 68112

Ripley, Ed & Irene, P.O. Box 9374, North St. Paul, MN, 55109

Rouette, Robert, 4000 Cardinal Leger - #14, Trois-Rivieres, QC, G8Y 2H2 Canada

Shields, Bud, 5200 Irvine - #488, Irvine, CA, 92720

Smith, Robert D., 3960 Estate Dr., Vacaville, CA, 95688

Stambaugh, Mike, 506 W. Springettsbury Ave., York, PA, 17403

Wilson, Gary, 7058 Sharp Rd., Swartz Creek, MI, 48473

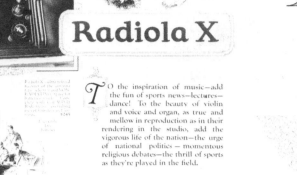

Radiola X

TO the inspiration of music—add the fun of sports news—lectures—dance! To the beauty of violin and voice and organ, as true and mellow in reproduction as in their rendering in the studio, add the vigorous life of the nation—the urge of national politics—momentous religious debates—the thrill of sports as they're played in the field.

New improvements for long distance receiving, selectivity, simplicity and perfect tone make the Radiola X an outstanding radio achievement of the year. And the same quality in *appearance* makes it the Radiola for the finest of city and country homes. It combines the joy of radio—and pride of possession!

"There's a Radiola for every purse"

Radio Corporation of America

Radiola

1924 Magazine Ad

Still Better
New Crosley Radio Receivers

THAT Crosley Radio Receivers have given complete satisfaction in the past is evidenced by the fact that, during the last twelve months, The Crosley Radio Corporation produced more receiving sets than any manufacturer in the world. That the new line of Crosley instruments, illustrated herewith, will give even better service is assured by the exhaustive tests to which each model has been subjected both in our laboratories and in actual use under all weather conditions.

Each Crosley Model is designed to give the utmost efficiency at the lowest cost. You may start with a small receiver and as you desire add to it to increase its range and volume.

Starting with the one tube Crosley 50 for $14.50 you can add the two stage Amplifier Crosley 50-A for $18.00 and have a three tube regenerative receiver for $32.50. Or, to the two tube Crosley 51 for $18.50, if greater range is wanted, add the one stage Amplifier Crosley 51-A for $14.00 thus making a three tube set for $32.50. Or purchase the Crosley 52 a three tube regenerative receiver for only $30.00.

You can pay much higher prices for radio receivers. But we have yet to find one at any price that will out-perform the Crosley Trirdyn 3R3 at $65.00 or set in special cabinet $75.00, the last word in Radio engineering.

For selectivity, ease of tuning and nicety of calibration, this instrument has astounded radio experts wherever it has been tried. The other Crosley Models will give comparatively equal performances. No matter which you choose, the clearest possible reception from exceptionally long distance is assured you.

Before you purchase a radio receiver listen in on a Crosley. Compare its performance with any other instrument on the market. We know then that you will choose a Crosley.

Crosley 50, $14.50

Crosley 51, $18.50

Crosley 51-P, $25.00

Crosley 52, $30.00

Crosley Trirdyn 3 R 3, $65.00
And Below
Crosley Trirdyn Special, $75.00

All Crosley Regenerative Receivers licensed under Armstrong U. S. Patent, 1,113,149

CROSLEY 50—A new one tube Armstrong Regenerative Receiver. We believe this to be the most efficient one tube receiver ever put on the market ....Price **$14.50**
Crosley 50-A, two tube amplifier may be added at $18.00

CROSLEY 51—Two tube regenerative receiver, the biggest selling radio receiver in the world. Gives loud speaker volume on local and distant stations under average conditions ....Price **$18.50**
Crosley 51-A, one tube amplifier may be added at $14.00

CROSLEY 52—A new three tube Armstrong Regenerative Receiver. Provides loud speaker volume on distant stations under practically all conditions ....Price **$30.00**

CROSLEY 51-P—This is our new portable set. It is the Crosley Model 51 two tube receiver mounted in a leatherette covered carrying case, battery space and all self-contained....................Price **$25.00**

CROSLEY TRIRDYN 3R3—This three tube receiver gives the efficiency and volume of five tubes. We believe it is the most efficient receiver on the market at any price for bringing in long distance stations ....Price **$65.00**

CROSLEY TRIRDYN 3R3 SPECIAL—The same as the Trirdyn 3R3 except cabinet is larger to contain "A" and "B" dry cell batteries and accessories. A beautiful set to match the highest grade of furniture....................................Price **$75.00**

THE CROSLEY RADIO CORPORATION
Powel Crosley, Jr., President

7314 Alfred Street   Cincinnati, Ohio

The Crosley Radio Corporation owns and operates broadcasting station WLW

CROSLEY
Better-Cost Less
Radio Products

1924 Catalog Page

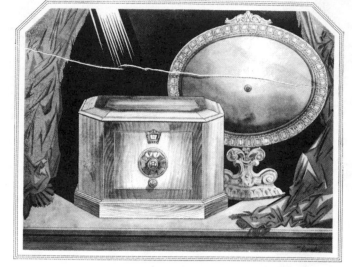

BARGAINS IN BATTERY SETS

We have listed here the greatest radio battery set bargains that have ever been offered.

The radio sets, which you will find listed in the following pages, are all standard sets made by the greatest and best known radio set manufacturers in America.

All of these sets are battery sets; but this is only one reason why they are sold at such ridiculously low prices.

The other reason is that these sets are mostly demonstration and display models from New York's largest radio and department stores.

We have been able to make connections with a number of houses in New York City, and we constantly secure these fine sets at remarkable prices. Due to these circumstances, we are enabled to sell the sets to you at only a fraction of their original cost.

Even if you do not need any one of these fine sets, here is a tremendous opportunity to get valuable radio parts at prices next to nothing.

Every one of these sets contains, in addition to the valuable cabinet, a goodly quantity of radio parts, such as condensers, transformers, sockets, tuning dials, coils, etc.

There is not a single set that does not contain anywhere from $15.00 to $50.00 worth of radio parts, if you had to buy them separately. In many cases, it will pay you to rip a set apart in order to get these valuable parts at unheard of prices. So, even if you do not need a set, the tremendous value in the separate parts will recompense you for the small outlay. AND FURTHER—

TURN THESE SETS INTO BIG MONEY!

There are still many families and many houses not equipped with radio today. At the prices at which we are selling them, it will pay you to install these sets and sell them at an excellent profit.

Most of the sets can readily be electrified by means of eliminators, or by making small structural changes, which anyone handy with tools and understanding radio can readily do. Most of the parts needed are listed in this catalog.

A number of our customers have made as much as $20.00 and $30.00 on each one of these sets by installing them, incidentally making a profit on tubes, loud speakers, eliminators, etc.

Remember, we do not sell you these sets as brand new. They all have been somewhat handled, but they are all in excellent condition, and, by going over the cabinets with some furniture polish, or otherwise renovating them, they will make a first class appearance and, in most instances, you will not be able to tell the set apart from a new one.

IMPORTANT!

Prices are for complete sets, as illustrated, including cabinets (Radiola 25 comes complete with loop). Prices include no accessories, such as tubes, loud speakers, etc.

In as much as it is not always possible to secure the exact set which you order, please be sure to fill out a second and third choice on the order blank below.

We will try very hard to ship you the exact set as ordered, but if it is impossible for the time being to secure it, WE WILL SHIP ONE OF EQUIVALENT VALUE, and, for this reason, we ask you to give us a second or third choice.

On this special sale, we cannot exchange sets nor accept returns on them for credit.

We have a good supply of these sets on hand, and, in most cases, can ship without delay.

Due to the exceedingly low prices, we cannot accept C. O. D. or part C. O. D. terms on these sets. Full amount must accompany every order.

Radiola Superheterodyne AR-812

One of the most famous radio sets in America. This set placed on a table, the battery switch turned to "on," and music will be heard—without an outdoor antenna; it works with a loop aerial built inside the cabinet. The set is super-sensitive and, in certain localities, it is possible, on the east coast, to hear west coast stations. The cabinet holds all the batteries for the six "dry-cell" tubes required. Some experimenters tune in short wave stations and use the AR-812 as the INTERMEDIATE FREQUENCY AMPLIFIER. In that way the tremendous amplification obtainable from this receiver is used to the fullest extent. A push-pull switch (center) turns the set on and off; another (lower left) cuts in either one or two stages of A.F. amplification. Although the cabinet is 35 inches long, 11½ deep and 11½ high, the panel of the receiver is only 19 inches long and 9 inches high. The difference lies in the two end compartments for "A" and "B" batteries. Six type UV199 tubes are required for this receiver. Drycell power tubes, the type '20, may be used in this set if a Naald or similar adapter is used. Shipping weight 45 lbs.

List Price is $220.00.

No. 2103 Radiola AR-812. YOUR PRICE **$10.95**

Atwater Kent 20 Compact

Five 201A tubes are used in this very sensitive and selective tuned radio frequency set. Dimensions: 20x6½x6½ inches high. A six-wire cable, "color coded," 6 ft. long, is included. Cabinet is finished in walnut. The panel is metal, finished in flat brown. Variable condensers having 16 plates are used. The variable condensers are independent of the receiver chassis. 3 brown molded "full vision" dials are used. A 3-point switch on the panel selects taps on the first R.F. coil, for "local" or "distance" reception. Non-oscillating. Easily re-wired for A.C. operation. Shipping weight 20 lbs.

List Price $60.00
No. 2100 A.K. Compact. YOUR SPECIAL PRICE. **$10.95**

Radiola 28 Superheterodyne

is a "Second Harmonic" super-heterodyne. However, the circuit of the "28" includes 7 type X199 tubes and 1 type X120 tube. The Radiola 28 includes 3 S.L.F. condensers, 2 1-mf. safety lamp bypass condensers; center-tapped loop (necessitated by the stage of neutralized R.F.); 1 off-on switch; 2 filament rheostats; 1 4-coil R.F. inductance; 2 jacks; and the special 8-socket "catacomb" containing the I.F., R.F. and A.F. transformers. Coast to coast reception is a rather usual accomplishment! Uses 2 drum dials—space for station logging thereon. Many easy ways of electrifying this receiver. Revolving loop—greatly assists tuning. Access to battery compartment obtained by raising receiver on hinge. A pressed steel frame supports the chassis equipment. The mahogany cabinet has the general appearance of a secretary. List Price $295.00.

No. 2106 Radiola 28.
YOUR SPECIAL PRICE **$24.98**

Atwater Kent 30 Compact

One of the best sets Atwater Kent produced. It is a six-tube set. Three stages of R.F. amplification, detector two stages of A.F. amplification. And single dial control. Even as an ordinary battery set it will sell on sight, as the glossy, molded bakelite "full vision" dial, volume controls, and highly gilded metal parts present an attractive appearance. There is a vernier wheel at the lower edge of the dial, for fine tuning. The variable condensers are 16-plate size. The two A.F. transformers are shielded. These transformers may be used as replacements as they will fit into almost every set. The sockets are demountable. Overall dimensions are 20x6½x6½ inches high. Shipping weight 20 lbs. A 6-wire color-coded cable 6 ft. long is included. 5 type 201A tubes and 1 type 112A or 171A are recommended for this receiver. List Price $80.00.

No. 2107 A.K. 30 Compact. YOUR SPECIAL PRICE **$14.95**

1931 Catalog Page

AMERICAN MICROPHONES—Microphones of Quality

Model CD

A two-button Microphone built especially for voice amplification, public address purposes. A two-button type made...

No. 9H3253. Model CD Microphone. List price $22.50. Dealer's price $11.50. Less 2%. Net **11.27**

Model FG Handy Case

This is a most popular type of case for use in remote control.

No. 9H3268. FG Handy Case. List price $6.5. **4.23**

Desk Stand Model BD

The Desk Stand contains a ring of...

No. 9H3269. **10.58**
No. 9H3270. **11.76**
No. 9H3279. **13.52**

Input Transformers

A Microphone Input Transformer is used in all Microphone hook-ups...

No. 9H3257. Input Transformer. List price $6.00. Dealer's price $5.43. Less 2%. Net **5.29**

MUTER Dependable "B" Socket Power Unit

It is a generally accepted fact, among those who really know the fine points of radio, that a Muter product can always be relied upon...

For 110-120 Volt 50-60 AC

No. 9H3013. **11.76**

Dumont Bone Dry "A" Eliminator

This Eliminator is of the highest grade throughout. Has no...

SPECIFICATIONS:

No. 9H3011—Dumont "A" Eliminator. List price $15.00 each. Dealer's price $12.00. Less 2%. Net **11.76**

1931 Catalog Page

SCREEN GRID BATTERY RADIO

7 Tube—Equipped with Batteries to Last One Year!

There are still vast territories where electric current has not as yet come into general use. Homes in these territories will gladly welcome this screen grid and battery radio because it combines all new developments in radio design heretofore confined to A.C. Sets, in a handsome, inexpensive battery radio...

One Year Service Without Battery Replacement

The new Eveready Air Cell "A" Battery furnished with this set is designed for these new two volt battery tubes...

Console

The Console is attractively yet simply designed. It is substantially constructed of genuine American Walnut with Maple overlays, hand rubbed to a velvety finish...

Comes Complete With Equipment

Set comes complete ready for installation with the following equipment:

1 Eveready Air Cell "A" Battery
3 45 volt Standard "B" Batteries
1 22½ volt "C" Battery
4 x33 Screen Grid RCA Licensed Tubes
1 x30 Power Detector RCA Licensed Tube
2 x33 Power RCA Licensed Tubes

9H1944 Screen Grid Battery Radio List $94.00, Dealers' Price $57.76 less 2%

56.21 NET

Complete With All Equipment

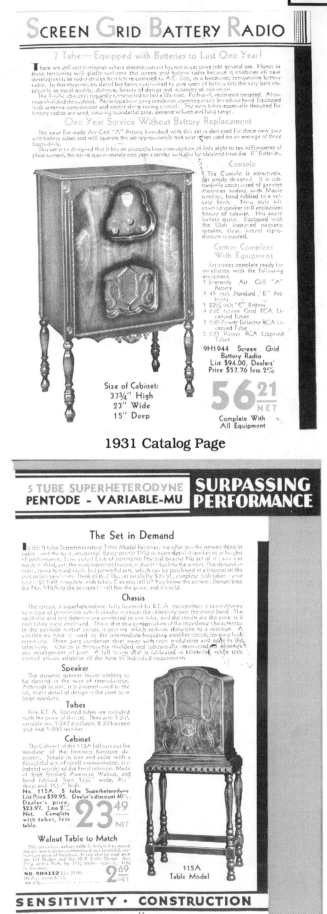

Size of Cabinet:
37¾" High
23" Wide
15" Deep

1931 Catalog Page

Sentinel RADIO

PORTROLA—AN UNUS 8 TUBES—VARIABLE-MU

New and Novel Cabinet Arrangement Assures Popularity of the Sentinel Portrola

Striking! Different! That's the new eight tube Sentinel Portrola. While yet a floor radio it can be moved from place to place with the ease of portability as a table model or midget...

A Brilliant Feat of Engineering

The same mechanical perfection that is characteristic of the entire Sentinel line is incorporated in the Portrola...

Cabinet Distinction

The cabinet, while novel in appearance, does not depend on novelty alone for its appeal...

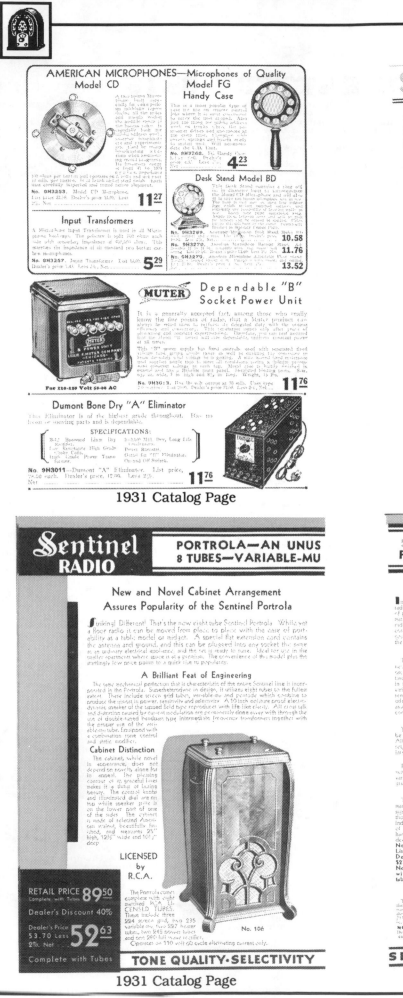

No. 106

LICENSED by R.C.A.

The Portrola comes complete with eight matched R.C.A. LICENSED TUBES. These include three 224 screen grid, two 235 variable-mu, two 227 heater type tubes, two 245 power tubes and one 280 full wave rectifier. Operates on 110 volt 60 cycle alternating current only.

RETAIL PRICE 89.50 Complete with Tubes
Dealer's Discount 40%
Dealer's Price 53.70 Less 2%. Net **52.63**
Complete with Tubes

TONE QUALITY·SELECTIVITY

1931 Catalog Page

---

5 TUBE SUPERHETERODYNE PENTODE - VARIABLE-MU

SURPASSING PERFORMANCE

The Set in Demand

In this 5 tube Superheterodyne Table Model Receiver, we offer you the newest thing in radio—and the most sensational...

Chassis

The circuit, a superheterodyne, fully licensed by R.C.A. Incorporated, is a revolutionary new type of production which greatly increases the selectivity over the entire band...

Speaker

The dynamic speaker leaves nothing to be desired in the way of reproduction...

Tubes

Five R.C.A. licensed tubes are included with the price of this set. They are: 3-235 variable-mu, 1-247 oscillator, 1-224 screen grid and 1-280 rectifier.

Cabinet

The Cabinet of the 115A follows out the inspiration of the foremost furniture designers. Simple in line and color...

No. 115A. 5 tube Superheterodyne List Price $39.95. Dealer's discount 40%. Dealer's price, $23.97. Less 2%. Net **23.49** Complete with tubes, less table.

Walnut Table to Match

No. 9H4112. **2.69 NET**

115A Table Model

SENSITIVITY · CONSTRUCTION

1931 Catalog Page

13

1932 Catalog Page

1932 Catalog Page

1932 Catalog Page

1932 Catalog Page

Kolster (6-D) 6 Tube Battery Receiver
In Console Less K-5 Speaker
Uses five 201-A, and one 112-A or 171-A Tubes

By popular demand, we can supply the famous Kolster 6-D battery-operated receiver in the Kolster 6-H console cabinet. This unit does not include the Kolster Electro-Dynamic Speaker and Power Amplifier. The speaker compartment has a baffle board for a 10" Speaker. The battery compartment has sufficient room for either batteries or eliminators. For outlying districts where A. C. or D. C. current is not available, this receiver will work exceptionally well. Just add any type of speaker to this set to make it complete.

$13.50
less tubes
Cat. No. S-2753

The new 2-volt tubes may be used with this receiver.

KOLSTER CABINET

Used with the 6-H Receiver and K-5 Dynamic Speaker. Can be used with any A. C. receiver chassis or can be built into a high class battery receiver.

Made of a beautiful burled walnut with maple overlay and full swinging doors. It takes a receiver front of 15 x 8½", has a battery compartment and a separate compartment with room for a 10" speaker with a baffle board provided.

$5.90 Size: 53 x 27 x 18¼
Cat. No. 1779

BREMER-TULLY
7 Tube 110 Volt, D. C. Receiver

1. D. C. All Electric
2. Seven Tubes in All.
3. Four Screen-Grid.
4. New Two-Volt Tubes.
5. Rola Dynamic Speaker.
6. Consumption: 17 Watts
7. Push-Pull Audio.

An entirely new creation to fill the demand for a D. C. Electro-Dynamic Receiver. It employs the latest type of shielded circuit using the new **Two-Volt Tubes; three Screen-Grid** Radio Frequency Stages, all tuned, using three UX232's; a tuned Screen-Grid Power Detector using a UX232; a quality first Audio Stage using a UX230 and a power Audio Stage using two UX231's in **Push-Pull.** The output is fed through a special output transformer, into a **Rola 7 inch Dynamic Speaker.**

The selectivity, distance getting qualities and general performance of this receiver is superb. It is sharp and does not require an elaborate antenna system. The cabinet is substantially built of grained walnut and stands 40 inches high, bringing tuning panel to eye level when sitting. **Less tubes**

Model 80 D. C.

$26.50
Cat. No. S-3453

New 1932 Universal I.C.A Companion Portable 7 Tube Radio

The I. C. A. COMPANION truly THE universal Radio . . . for use in the city . . . country . . on the farm . . . in automobile . . . boat or airplane . . . anywhere.

Uses two 233, three 237, two 236 tubes

Operates on either A. C., D. C. or with batteries. Weighs only 14 lbs. 7 Tubes, Volume Control, Built-in Speaker, Single Dial Control, Leatherette case.

The I. C. A. Universal Companion answers the world-wide demand for a single radio, usable on either A. C. or D. C. electric current (105 — 125 volts) and also on battery power (6 volt storage A battery or dry A battery and 135 volts B battery).

Just think of it—a single radio receiver that takes the place of three radio sets.

The circuit is the last word in radio using the new 6.3 volt, screen grid heater tubes, allowing considerable voltage variation without affecting reception.

Change over from electric A. C. or D. C. to battery operation is accomplished by merely inserting a plug which is furnished with each receiver into either one of the sockets marked for electric and for battery. No change of tubes.

Only a short antenna or ground connection necessary for perfect reception. A special jack provided for head phone use. All self-contained in handsome leatherette traveling case measuring 12 x 11 x 8½.

$23.70
(less tubes)
Cat. No. 1407

National Phanstiehl Screen-Grid 5 Tube Midget Receiver

Subject to standard R.M.A. guarantee

Uses 201A amplifier, 224 screen-grid T. R. F. amplifier, 227 power detector, 112A power audio stage and a 280 full wave rectifier.

Absolutely the smallest, lightest and cheapest midget receiver made today. It enables every home to have a radio. It uses the latest type of Utah Magnetic Speaker which reproduces faithfully.

Will receive distant signals as well as the powerful and weak local stations. Does not need a large antenna due to the fact that it employs screen-grid R. F. amplification. It weighs but 9 pounds.

Size: 9¾ x 13½ x 7½"

$11.50
less tubes
Cat. No. S-1963

American Crystal Receivers

These inexpensive, highly efficient, miniature receivers open an entirely new field of profit to the radio dealer. There is practically no limit to the number of these sets that can be sold. There is a type for every age and requirement. The sets come equipped with Supersensitive Crystals and require no bathterfies, therefore costing nothing to operate. It is the logical choice)for the smaller boy. Boost your sales with these receivers, they are as popular as ever.

THE "SUPERTONE" SET
Has a range of from 200 to 600 meters and will receive signals further than 25 miles. Fixed detector is dust-proof.

75c
Cat. No. 1957

Complete with Head Phones and Antenna Equipment, ready to operate.

$2.25
Cat. No. 1958,

THE "SELECTIVE" SET
This is the finest of all crystal receivers. Has a variable condenser and will bring in one station at a time. Loud and clear.

$2.50
Cat. No. 1959

Complete with Head Phones and Antenna Equipment, ready to operate.

$3.95
Cat. No. 1960,

Pandora Crystal Set

Selective. Very efficient. Receives loud and clear when operated within twenty-five miles of broadcasting stations. Metal beautifully enameled in various colors. With crystal.

$1.15
Cat. No. E-62

Complete with Head Phones and Antenna equipment, ready to operate. Cat. No. E-62C

$2.65

For 220 Volts, A. C.
For 220 volts A. C. use step-down transformer consuming approximately 50 watts. Cat. No. 1408

$3.15
Cat. No. 1408

For 220 Volts D. C.
For 220 volts D. C. use external resistance housed in a perforated metal box. Cat. No. 1409

$2.70
Cat. No. 1409

1932 Catalog Page

16

BIG RADIO PARADE!
for AUTO...HOME...VACATION

"His Master's Voice" at your fingertips 24 hours a day
... wherever you are, whenever you want pleasure—and all at little cost!

ON SALE AT ALL RCA VICTOR DEALERS

$19.95

$21.50

$24.95

VICTOR RECORDS
for "2 in One Music"

Model RE-40
"Radio Femograf"

$49.95

$44.50

$49.95

MAIL COUPON FOR FREE GIFT!

1933 Magazine Ad

Brand New..each a PHILCO
..each *supreme* in its field

THEY vary in price — they vary in size — but they are all one in quality. For all are PHILCOS — all true representatives of the name that ranks highest in radio.

Whatever you seek in radio — PHILCO will supply. If you desire the truest reproduction of programs from far and near — PHILCO suggests those great musical instruments — the X Models with the Patented Inclined Sounding Board.

If you prefer a radio of smaller size — possibly an exquisite lowboy — you will find a PHILCO to meet your requirements. If you prefer a still smaller set — the Baby Grand models give you tremendous power and distance range. Even the PHILCO Compact is a radio of true PHILCO quality — if you want a set to carry on your travels or as an extra radio at home.

From a tiny set you can tuck into your suitcase to a powerful radio that will flood an auditorium with melody — at every step there is a PHILCO definitely designed to serve a particular purpose — and supreme in its particular field. All await your inspection and selection at the nearest PHILCO dealer.

A musical instrument of quality

PHILCO 14L — $85 ▶

PHILCO 16X — $150

◀ PHILCO 54C — $25

PHILCO REPLACEMENT TUBES IMPROVE THE PERFORMANCE OF ANY SET

$15 to $250

1933 Magazine Ad

ATWATER KENT RADIO

**NOT DAYS...
NOT MONTHS...
BUT YEARS *of*
happiness...**

1933 Magazine Ad

Majestic Royale

MODERN CHARM—A HINT OF CHINESE CHIPPENDALE—AND HALF THE PRICE YOU'RE GUESSING!

Majestic Master Six

JUST AS SMART AS "SMART" CAN BE

"Smart Set" of Radio

Majestic Century Six

AS KEEN AND CAPABLE AS A NEW SPORT-MODEL ROADSTER

Majestic Twin-Six

THREE'S COMPANY! THE MAJESTIC TWIN-SIX CAR RADIO—AND YOU

Majestic Radios are Standard Equipment on all Essex Terraplane De Luxe Models

The **SMART SET**
Majestic
RADIO

1933 Magazine Ad

1935 Catalog Page

1935 Catalog Page

1935 Catalog Page

1935 Catalog Page

17

1936 Catalog Page

1936 Catalog Page

1936 Catalog Page

19

Sentinel 6-Volt BATTERY RADIOS
Models
No B or C Batteries

7-TUBE ALL-WAVE SUPERHETERODYNE

● 3 Bands—16.5 to 555 Meters ● New Oversize Illuminated Airplane Dial ● Synchronous Vibrator—No B or C Batteries Necessary ● Automatic Volume Control ● Tone Control ●

A new type of farm radio which completely eliminates the use of B and C batteries and forever abolishes their expense and nuisance. The only power supply needed is the conventional 6 volt storage A battery. Actual operating expense should not exceed 1c a day. Will positively get foreign stations under practically any condition as well as American short wave and regular broadcasts. Tunes 16.5 to 52; 52 to 170; 175 to 555 meters. Has a beautiful black and silver oversize dial, with individual wave band color illumination. Superheterodyne circuit uses the following tubes: 1-1C6, 2-34, 3-30, 1-19. Class B Amplification is employed. Automatic volume control and tone control further enhance performance. Extremely low battery drain—only 1.55 amps.

5-TUBE SUPERHETERODYNE
Short Wave and Regular Broadcasts

Just look at these features—synchronous vibrator (no B or C Batteries required)—2-band tuning—illuminated airplane dial—tone control—automatic volume control. Tunes from 120 to 130 and from 175 to 500 meters, covering regular broadcast and all police calls. New oversize black and silver dial, illuminates each band separately; red for regular broadcast, green for short wave. Low battery drain. Uses the following tubes: 1-6C6, 1-34, 1-6B7, 1-30, 1-19. 6" magnetic speaker. Cabinet of selected walnut veneers and maple overlays. 16" high, 12½" wide, 8½" deep. Weight, 22 lbs.

No. 9H4235. Sentinel 5-tube 6-volt Table Model. Complete with RCA Radiotron tubes, less battery. List $39.95. Dlr's., ea., $25.97, less 2%, net.. **25⁴⁵**

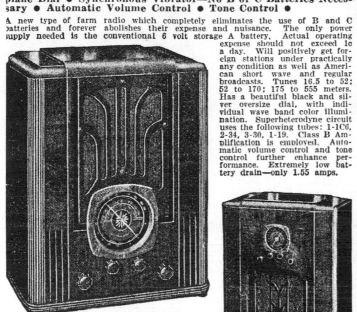

TABLE MODEL

Cabinet is of a smart, modern design that has strong universal appeal. Has a genuine walnut front of selected veneers, the center section of panel being sliced walnut with pin stripe overlay side panels. Dimensions: 17" high, 13" wide, 9½" deep. Weight, 25 lbs. Equipped with 6" speaker.

No. 9H4236. Sentinel 7-tube 6-volt table model, complete with RCA Radiotrons, less battery. List, $49.95. Dlr's., ea., $32.47, less 2%, net............. **31⁸²**

CONSOLE MODEL

A handsome console cabinet with butt walnut instrument panel, horizontally striped walnut front panels, front walnut top and overlay. Rounded top and corners. Equipped with 8" speaker. Dimensions 39" high, 24" wide, 13" deep. Will hold battery. Weight, 62 lbs.

No. 9H4237. Sentinel 7-tube 6-volt Console, complete with RCA Radiotron tubes, less battery. List, $69.95. Dlr's., each, $41.97, less 2%, net........ **41¹³**

Refer to Index for Listing on 6-Volt Batteries and Battery Chargers.

Sentinel 32-Volt
All Electric Receivers

6-TUBE SUPERHETERODYNE
Operates Direct from 32-Volt Lighting Plant

● American and Foreign Short Wave ● Illuminated Airplane Dial ● Automatic Volume Control ● Variable Tone Control ●

Here is the ideal receiver for those millions of homes and farms equipped with 32 volt farm lighting plant. Operates directly from lighting plant without the use of troublesome vibrators, dynamotors or any other type of "B" supply. Tunes from 47.5 to 130 and 175 to 550 meters, providing reception of the entire broadcast band and the better Foreign Short Wave, domestic, all police, airplane and amateur signals. The 6-tube superheterodyne circuit incorporates such features as automatic volume control, tone control, large illuminated airplane dial, and many other refinements. A super-sensitive electro dynamic speaker provides perfected tone. Draws only 50 watts—less than an ordinary light bulb. Tubes used are as follows: 1-6A7, 1-6D6, 1-75, 1-76, 2-48. Push-pull type amplification is employed.

TABLE MODEL

Semi-modernistic cabinet of selected walnut. 16" high, 12" wide, 8½" deep. Wgt., 17 lbs.

No. 9H4238. Sentinel 32 volt Table Model, complete with RCA Radiotron tubes. List $44.95. Dlr's., ea., $25.97, less 2%. net. **25⁴⁵**

CONSOLE

Substantially constructed console has beautiful butt walnut and streamline top construction. Attractive routings on front. Hand-rubbed piano finish. Dimensions: 37" high, 22" wide, 12" deep. Wgt., 55 lbs.

No. 9H4239. Sentinel 32 volt Console, complete with RCA Radiotron tubes. List $59.95. Dlr's., ea., $35.97, less 2%, net................... **35²⁵**

1936 Catalog Page

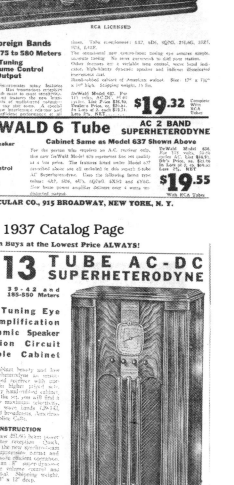

CAMDEN 12 TUBE AC-DC
2 BAND SUPERHETERODYNE
"Super 12"

RANGE—39-142 and 185-550 Meters
AMERICAN — FOREIGN — AMATEUR
Police - Aviation-Ships-at-Sea—Etc.

RCA LICENSED

Beam Power Amplification

- Automatic Volume Control
- Illuminated Airplane Dial
- "Phantom" Shielded Aerial System
- Full Range Dynamic Speaker
- Beautiful Deluxe Lowboy Cabinet
- New Beam Power Output Tube

Superb Tone——DeLuxe Performance

MODEL 12S9.
List Price $34.95
Dealer's Price, Each $15.95
Less 2%, NET ....

$15.63
COMPLETE WITH RCA LICENSED TUBES

RADIO CIRCULAR CO., 915 BROADWAY, NEW YORK, N. Y.

1937 Catalog Page

Latest 1938 Model 637
DeWALD 7 Tube AC-DC
2 Band Superheterodyne

RCA LICENSED

American and Foreign Bands
RANGE: 16 to 53——175 to 580 Meters

- Syncro-Beam Tuning
- Automatic Volume Control
- Beam Power Output

DeWald Model 637. For
115 volts, AC-DC, 40-60
cycles. List Price $36.50.
Dealer's Price, ea. $20-54.
In Lots of 2, each $19.73)
Less 2%, NET ....

$19.32
Complete With RCA Tubes

New 1938 Model 636 DeWALD 6 Tube
AC 2 BAND SUPERHETERODYNE
Cabinet Same as Model 637 Shown Above

- Powerful Dynamic Speaker
- Syncro-Beam Tuning
- Wave Band Indicator
- Automatic Volume Control
- Over 4 Watts Output
- Variable Tone Control
- Fine Deluxe Cabinet

DeWald Model 636.
For 115 volts, AC. List $33.95.
Dlr's. Price, ea. $20.79.
In Lots of 2, ea. $19.95
Less 2%, NET

$19.55
With RCA Tubes

RADIO CIRCULAR CO., 915 BROADWAY, NEW YORK, N. Y.

1937 Catalog Page

NEW 1937 Sensational Value! Camden 5 TUBE
RCA LICENSED

DUAL WAVE 66.5-200 AND 200-560 METERS AC-DC COMPACT

CAMDEN MODEL 5TD, 5 TUBE A.C.-C.E.

List Price $21.50.
OUR PRICE, EACH $12.75
In lots of 2, $11.95 ea.
In lots of 5, ea. ...... ss.

$11.10
COMPLETE WITH RCA RADIOTRONS

Camden 6 TUBE
RCA LICENSED
with Dynamic Speaker
A. C. - D. C.
COMPACT
LONG AND SHORT WAVE
WAVE RANGE 42-185, 200 to 560 METERS

CAMDEN MODEL No. 6TD A.C.-D.C. COMPACT

List Price $29.75
OUR PRICE $16.85
EACH
Lots of 2, $15.95 ea.
Lots of 6, ea. ......

$15.25
COMPLETE WITH RCA RADIOTRONS

RADIO CIRCULAR CO., 915 BROADWAY, NEW YORK, N. Y.

1937 Catalog Page

New 1938 DeWALD 13 TUBE AC-DC SUPERHETERODYNE
2 BANDS 39-42 and 185-550 Meters

- Syncro-Beam Tuning Eye
- Beam Power Amplification
- Powerful 8" Dynamic Speaker
- Noise Suppression Circuit
- Beautiful Console Cabinet

DeWald Model 1300-C. For
115 volts, AC-DC, 40-60 cycles.
List Price $42.75.
Dealer's Price, Each $35.77.
In Lots of 2, Each, $33.89.
Less 2%, NET ....

$33.21
With RCA Licensed Tubes

RCA LICENSED

New 1938 DeWALD "MOTORTONE"
5 TUBE SUPERHETERODYNE
AUTO RADIO
RCA LICENSED
FULL-SIZED RADIO PERFORMANCE AT A SENSATIONALLY LOW PRICE

FEATURING:
- Dynamic Speaker
- Automatic Volume Control
- Single Hole Dash Mounting

Streamlined Model 527

DeWald Model 527.
6A7, 6D6, 75, 41, 84. List Price $29.95.
DEALER'S PRICE, Each $17.97,
IN LOTS OF 2, EACH $16.15 . Less 2%, NET ...

$15.85
With RCA Licensed Tubes

RADIO CIRCULAR CO., 915 BROADWAY, NEW YORK, N. Y.

1937 Catalog Page

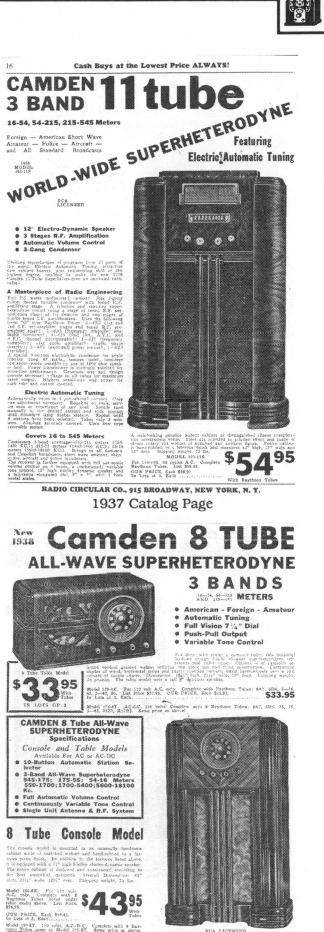
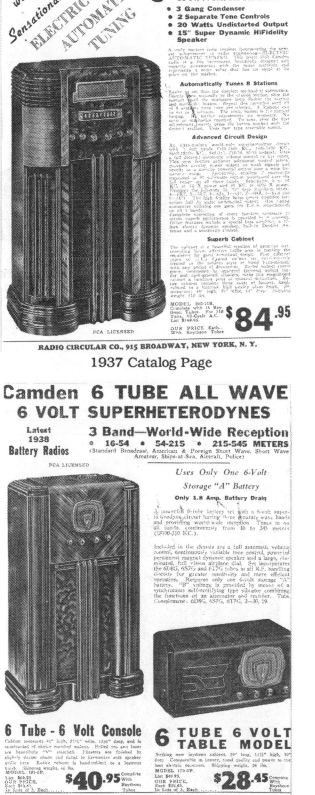

1938 Catalog Page

1938 Catalog Page

1938 Catalog Page

1938 Catalog Page

23

NEW 7-TUBE A.C. ELECTRIC MANTEL RADIO $25.95 Cash

$3 Down, $4 Month

Attractive Table Model of powerful 7-tube Radio shown and described below. *Hand Rubbed* to a gleaming finish! Walnut Veneer top, front and sides, lovely figured Walnut center panel, and beautifully grained bands. *6-inch Super-Dynamic Speaker.* Easy to buy on terms. Complete with Tubes and Instructions. Aerial extra, see Page 665. Size: 16¼ by 9 by 14¼ in. high. For use on 105 to 125 volt, 50 to 60 cycle, A.C. only. *Shipped from* Baltimore, Albany or Pittsburgh. Send order to nearest House. (25-cycle Model $2.50 extra and shipped from Chicago Factory stock.)

P462 C 715—Mantel Model. Ship. wt. 25 lbs. *Mailable.* Cash Price.....................$25.95
Time Payment Price: $3 Down, $4 a Month....................................$28.25

POWERFUL 8-TUBE A.C. ELECTRIC MANTEL RADIO $34.95 Cash

$4 Down, $5 Month

Beautiful Table Model of distance-getting Console Grande Radio described below. *Quality, Performance and Beauty* equal to radios selling as high as $69. Big 8-inch Super-Dynamic Speaker covers practically the entire musical range with astounding realism. Cabinet is sliced Walnut Veneer with horizontal bands. Enjoy it while paying for it. Complete with Tubes and Instructions. Aerial extra, see Page 665. Size: 22 by 10¾ by 12 in. high. For use on 105 to 125 volt, 50 to 60 cycle A.C. only. *Shipped from* Baltimore, Albany or Pittsburgh. Send order to nearest House. (25-cycle model $2.50 extra and shipped from Chicago Factory.)

P462 C 800—Mantel Model. Ship. wt. 35 lbs. *Mailable.* Cash Price.....................$34.95
Time Payment Price: $4 Down, $5 a Month....................................$38.05

7-TUBE A.C. ELECTRIC CONSOLE RADIO $35.95 Cash

$4 Down, $5 a Month

Powerful 7-Tube, Super Heterodyne Circuit that brings you greater distance—your choice of the finest Foreign and American programs; Police calls from many cities—and almost unbelievable beauty and realism of tone. Two wave bands, 540 to 1730 K.C. and 5.7 to 18.0 M.C.

Personal Tone and Volume Controls. Turn Tone Control down to emphasize the mellow bass —up if you prefer the brilliant treble. Vary the volume from a light whisper to full orchestra intensity with Volume Selector; Automatic Volume Control keeps it there. New lighted Drum dial; Flush Controls make tuning easier. Automatic tuning on your six favorite stations.

Mighty 8-inch, Super-Dynamic Speaker with 12-inch Projectotone that assures even greater realism of tone—plenty of volume. Electric Tuning Eye assures best reception.

Beauty to thrill the most ardent furniture lover. Sides, top, and dial panel are made of fine figured Walnut Veneer with bands of genuine Mahogany as a stunning contrast.

Complete with Tubes and Instructions. Aerial extra, see Page 665. Size: 24¼ by 12¼ by 40 in. high. For use on 105 to 125 volt, 50 to 60 cycle, A.C. only. *Shipped from* Baltimore, Albany or Pittsburgh. Send order to nearest House. (25-cycle Model $2.50 extra and shipped from Factory stock in Chicago). Enjoy this fine radio now, pay while using it. See Page 1009.

P162 C 714—Console Model. Ship. wt. 58 lbs. *Not Mailable.* Cash Price.........$25.95
Time Payment Price: $4 Down, $5 a Month............$39.15

652 WARDS BA

8-TUBE CONSOLE GRANDE, ONLY $5 A MONTH $43.95 Cash

$5 Down, $5 A Month

Far Reaching Super Heterodyne Power makes distant stations seem only a few miles away. Incredibly life-like tone from its large 8-inch Super-Dynamic Speaker with 12-inch Projectotone—almost like being in the radio studio.

Foreign and American Wave Bands (police calls from many cities) provide a front row seat for the World's finest entertainment. Range 540 to 1730 K.C. and 5.7 to 18.0 M.C. With convenient 6-station Automatic Tuning. Automatic Volume Control that keeps Volume constant.

Automatic Bass Booster for marvelous beauty of tone. Turn the volume up or down. You'll find the proper balance between the bass and treble is always maintained. Keeps bass notes audible even when radio is playing very softly. Personal Tone and Volume Controls, and many other features.

Large Lighted Drum Dial and new, exclusive Flush Controls (for dial and volume)—far simpler to operate. Numbers and call letters are larger and easier to read . . . makes tuning more accurate.

Latest Console Grande Style Cabinet made of finest woods. Top and sides of sliced Walnut Veneer; dial panel of stump Walnut Veneer. Electric Tuning Eye assures more accurate reception.

Complete with Tubes and Instructions. Aerial extra, see Page 665. Size: 28 by 12¼ by 35 in. high. For use on 105 to 125 volt, 50 to 60 cycle, A.C. only. *Shipped from* Baltimore, Albany or Pittsburgh. Send order to nearest House. (25-cycle model $3 extra and shipped from Chicago Factory stock.)

P162 C 801—Console Grande Model. Ship. wt. 69 lbs. *Not Mailable.* Cash Price.........$43.95
Time Payment Price: $5 Down, $5 a Month (Include a new Aerial, see Page 1009 for terms) $47.85

1939 Catalog Page

Sonora-Model M-22
4 TUBE - 6 VOLT SUPERHET

- 8 Tube Performance
- Beam Power Output
- Illuminated Slide Rule Dial
- Synchronous Vibrator
- Automatic Volume Control
- Low Battery Drain
- No "B" or "C" Batteries
- Striking Bakelite Cabinet

$15.28
Net
In Lots of 3

An exceptional performer, equal in every detail to ordinary 8 tube receivers. Employs a superheterodyne circuit of advanced design unusually sensitive and selective. The use of dual and triple purpose tubes, plus the dual purpose synchronous vibrator, gives this set the power and range of the largest battery receivers. Tunes from 535 to 1720 K.C.'s (includes 1712 K.C. police channel) on a full vision illuminated slide rule dial. Requires only one 6 volt storage "A" battery. No "B" or "C" batteries. Unusually low drain of 1.9 amps.—actually 50% less than that consumed by other battery receivers of equal operating efficiency. Special permanent magnet dynamic speaker requires no battery current. Complete with the following tubes: 1—6A8G, 1—6K7, 1—6Q7G, 1—6G6G. Gorgeous two-tone bakelite cabinet in fluted design with special honey-comb-effect grille. Front in ivory; sides and top in black. 11¾" long, 7½" deep, 7¼" high. Wt. 11½ lbs. **No. 9H4286.** List $24.95. Dlr's., ea. $16.11. Lots of 3, ea. $15.59, less 2%, net.... **15.28**

6 Station Automatic Push Button Tuning!

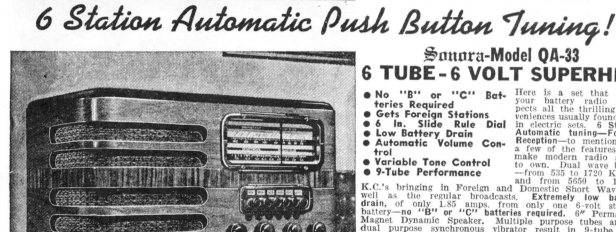

Sonora-Model QA-33
6 TUBE - 6 VOLT SUPERHET

- No "B" or "C" Batteries Required
- Gets Foreign Stations
- 6 In. Slide Rule Dial
- Low Battery Drain
- Automatic Volume Control
- Variable Tone Control
- 9-Tube Performance

Here is a set that offers your battery radio prospects all the thrilling conveniences usually found only in electric sets. 6 Station Automatic tuning—Foreign Reception—to mention but a few of the features that make modern radio a joy to own. Dual wave bands —from 535 to 1720 K.C.'s, and from 5650 to 18,000 K.C.'s bringing in Foreign and Domestic Short Wave, as well as the regular broadcasts. **Extremely low battery drain,** of only 1.85 amps. from only one 6-volt storage battery—no "B" or "C" batteries required. 6" Permanent Magnet Dynamic Speaker. Multiple purpose tubes and a dual purpose synchronous vibrator result in 9-tube performance. 2 watt output. **Full A. V. C.** and continuously variable tone control. Complete with the following tubes: 1—6C8G, 1—6H4G, 1—6F7G, 1—6G7G, 1—6L5G, 1—1J6G. Smartly styled cabinet in latest 1939 design with attractive corner type grill. Horizontal overlay trim of vertically grained walnut striping crosses the front face. Instrument panel is of quarter-cut choicely grained walnut. 17½" long, 8" deep, 7¼" high. Wt. 22 lbs. **No. 9H4105.** List $49.95. Dlr's., ea. $28.00. Lots of 3, ea. $26.65, less 2%, net **26.12**

3 BANDS - *World-Wide Reception!*
Sonora-Model 170-6B
6 TUBE - 6 VOLT SUPERHET

An unusual value in a precision built 6-volt Receiver. 3 full bands provide a world wide range of reception. Superbly designed Superheterodyne circuit. Employs a synchronous, self-rectifying vibrator which combines the functions of an alternator and rectifier, thereby obtaining the equivalent of 7-tube performance. Equipped with full automatic volume control and continuously variable tone control. Undistorted output of 2.1 watts. Low battery drain of 1.8 amps. Full vision airplane type dial. Complete with R.C.A. tubes as follows: 1—6D8G, 1—6S7G, 1—6T7G, 2—30, 1—19. Handsome laydown type cabinet with top, front and bottom made from single piece of richly grained walnut. Rolled effect top and bottom. Hand rubbed and highly polished. Dimensions: 19" long, 11½" wide, 10" deep. Wt. 24 lbs. **No. 9H4243.** List $49.95. Dlr's., ea. $22.50. Lots of 3, ea. $21.25, less 2%, net **20.82**

1939 Catalog Page

1940 Catalog Page

1940 Catalog Page

1940 Catalog Page

1940 Catalog Page

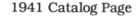

J 39.95   **K** 34.95   **F** 32.95   **H** 54.95   **G** 24.95

Nationally-Known General Electric Table Radios

CHOOSE FROM G.E.'S MOST POPULAR MODELS IN CLOCK, FM-AM AND REGULAR RADIOS

ATTRACTIVE TABLE RADIO 32.95

F This decoratively styled table set is a fine, popular model with more selectivity and sensitivity than (G) at right. It is compactly designed and offers the most in fine listening pleasure at a low price. It's a handsome model that goes well in any surroundings. It's good for outlying areas where distance-getting ability is required.

RED MAHOGANY OR IVORY COLORED PLASTIC CABINET with vertical grille louvers and decorative trim. Large cabinet size makes good tone chamber. Large slide rule dial is "spotlighted" by dial beam as pointer tunes stations.

5 TUBES PLUS RECTIFIER in dependable chassis. Uses tuned r.f. stage for the extra sensitivity and selectivity; 3-gang condenser with high ratio tuning. Has plenty of power for picking up distant stations when used with outside antenna—see antennas on Page 705.

DYNAPOWER SPEAKER has good, clear tone. Built-in Beamascope antenna for good station pickup. Has all other G.E. radio features listed at right below. Size of cabinet 12¹⁵⁄₁₆ in. wide by 8¾ in. high by 7¼ in. deep. For 105- to 125-volt, DC or 60-cycle AC. *Postpaid.*

62CP622M—Red Mahogany .$3.50 Down, $5 Mo. or Cash $32.95
62CP623M—Ivory. $3.50 Down, $5 Mo......or Cash 32.95

LOW-PRICED G.E. RADIO 24.95

G Select this outstanding Table Radio in choice of your favorite color to match your kitchen or bedroom furnishings. Although low in price, these fine radios are built to the same high standards of quality in design and appearance as all other G.E. radios listed on this page. It makes an ideal extra radio to have around the house for carrying from room to room to use, or for the children to hear their own programs.

PLASTIC CABINET IS SMARTLY STYLED IN BROWN MAHOGANY, IVORY, OR LIGHT MAROON. Extra large dial has large, easily read station numerals which are "spotlighted" by dial beam as pointer tunes stations.

4 TUBES PLUS RECTIFIER in dependable chassis built to last for years of service. Gives excellent reception in areas that are close-in to radio stations.

DYNAPOWER SPEAKER provides good tone and ample volume to fill an average size room. Has all the other General Electric radio features listed at right below. Size 12¾ inches wide by 6 inches high by 6½ inches deep. For 105- to 120-volt, DC or 60-cycle AC.

62 CP 614 M—Brown Mahogany color. *Postpaid*......$24.95
62 CP 615 M—Ivory color. *Postpaid*................. 24.95
62 CP 616 M—Light Maroon color. *Postpaid*......... 24.95

POWERFUL FM-AM TABLE RADIO 54.95

H General Electric's finest Plastic Table Radio. Tunes both FM and AM—two bands for double listening enjoyment with a greater variety of programs. Gives really powerful performance on both bands. Good distant reception on AM, and noise-free, life-like reproduction of music on FM. Built to give years of fine performance.

BROWN MAHOGANY COLORED PLASTIC CABINET has unique and striking appearance. "Sunburst" plastic dial with etched FM and AM scales is framed by highly polished beveled band, and is indirectly illuminated by dial light.

7 TUBES PLUS SELENIUM RECTIFIER in exceptionally fine chassis. Expertly designed FM circuit with high quality components—gives freedom from static and true musical reproduction. Has tuned r.f. stage for increased sensitivity and selectivity; 3-gang tuning condenser on AM.

G.E. DYNAPOWER 5¼-inch SPEAKER gives clear and natural tone. Full range tone control inside tuning knob. Powerline antenna for FM reception, built-in AM Beamascope antenna. Set incorporates all other G.E. features listed below. Size 13½ in. wide by 8¹⁄₁₆ in. high by 7⁹⁄₃₂ in. deep. For 105- to 120-volt, DC or 60-cycle AC.

62 CP 609 R—*Prepaid*—$14.50 Down, $5 Mo....or Cash $54.95

BEST ELECTRONIC SERVANT RADIO 39.95

J An excellent Clock Radio with "electronic servant" that does all these things: Turns on any appliance using less than 1100 watts that is plugged into receptacle on radio. Turns on radio automatically at pre-set time for waking you up to music. Has "wake-up" switch with buzzer alarm for arousing the sound sleeper. Buzzer will sound about 7 minutes after radio turns on, if desired. Slumber switch shuts off radio or appliance at any time during 60-minute period. Use it as a radio in the bedroom for listening to your favorite programs before retiring and as a clock at all times. Has accurate self-starting, self-regulating G.E. Clock.

CABINET IN RICH BROWN MAHOGANY (CORDOVAN) PLASTIC FINISH. Powerful chassis has 5 tubes plus rectifier, including tuned r.f. stage for greater sensitivity and selectivity. Powerful pickup on distant stations with outside antenna; see Antennas on Page 705. Clock operates while cord is plugged into outlet. Has G.E. radio features listed at right, including Beamascope antenna, Dynapower speaker. 11¼ in. wide, 5¾ in. high, 5 in. deep. For 105- to 125-volt, 60-cycle AC.

62 CP 665 M—*Postpaid*—$4 Down, $5 Mo......or Cash $39.95

ELECTRONIC SERVANT RADIO 34.95

K This fine Clock Radio will wake you up in the morning to the sound of a buzzer or will automatically turn on your favorite radio program at the time it is set for. It automatically starts coffee-maker, sunlamp, or any other electrical appliance up to 1100 watts at the time you desire—and will turn it off automatically at any time within a 60-minute period. You can leave the house and let this electronic servant turn on a fan, air-conditioner, etc., at any time within 24 hours before you return. Provides accurate time with self-starting, self-regulating G.E. clock.

BEAUTIFUL PLASTIC CABINET in choice of Ivory, Medium Red or Maroon color finish. Fine radio chassis has 4 tubes plus rectifier. Clock operates all the while cord is plugged into outlet; radio turns on by clock or by switch. Has all General Electric radio features listed at right. For 105- to 125-volt, 60-cycle AC. Size 11⁵⁄₁₆ inches wide by 6³⁄₁₆ inches high by 5¹⁄₁₆ inches deep. *Postpaid.*

62 CP 655 M—Maroon finish. $3.50 Down....or Cash $34.95
62 CP 656 M—Ivory finish. $3.50 Down....or Cash 34.95
62 CP 657 M—Medium Red finish. $3.50 Down.or Cash 34.95

FEATURES OF GENERAL ELECTRIC RADIOS

The fine-performing General Electric Radios listed on this page are representative of the complete General Electric Radio line. These latest models are smartly styled to look well in any room and yet they are moderately priced to make it easy for anyone to afford one. All General Electric radios have modern Superhet circuits which are time-tested designs that have been improved to a point of maximum efficiency and dependability.

They tune Standard Broadcast 540 to 1600 KC for regular radio programs, and all have built-in Beamascope antennas, an exclusive General Electric design. The FM-AM model also tunes FM band 88 to 108 MC and has additional antenna built in for FM reception. High quality Dynapower speakers with Alnico 5 magnets give more power and finer tone. All sets have full range volume controls. Clock radios use the nationally-known self-starting, self-regulating General Electric clock which operates all the while the radio is plugged into outlet. All General Electric radios are approved by Underwriters' Laboratories for safety in design and construction. For better distance reception, an outside antenna is recommended—see Page 705.

1951 Catalog Page

A.C. Cossor Ltd., Model #499 English
1940s
Ira & Debbie Grossman

A.C. Dayton, Model #XL-5
1925
Greg Nelson

Addison, Canada
1935
Michael Feldt

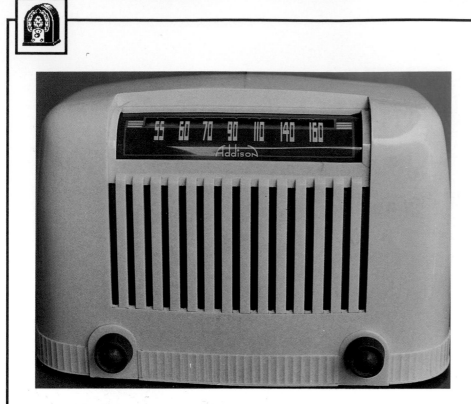

Addison, Model #55
1949
Robert Rouette

Admiral, Model #52-6A
1952
Don Nordboe

Admiral, Model #6T02
1946
Spencer Doggett

Admiral, AC Spark Plug
1962
Joe Ricci

Aetna
1936
Jay Kinnard

Air Castle, Model #320
1947
Richard's Radio of Omaha

Air Castle, Model #831
Jay Daveler

Airline, Model #62-476
1941
Don Nordboe

Air King, Model #4602
1945
Jay Daveler

Airline
1970s
Don Nordboe

Airline, Model #04BR-511A
1946
Ira & Debbie Grossman

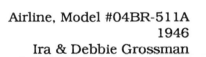

Airline, Model #25-WG-1573A
1953
Doug Burskey

Airline, Model #64BR-1808A
1947
Doug Burskey

Airline, Model #84BR-1502A
1946
Megan Amos

Airline, Model #84BR-1520B
1949
Spencer Doggett

Airline, Model #84GCB-1062A
1948
Don Nordboe

Airline, Model #93BR-421B
1939
Ira & Debbie Grossman

Airline, Model #DC-6
1925
Greg Nelson

Airline, Model #93WC-604
1939
Jay Daveler

Aladyne, Model #RF4
1925
Michael Feldt

All American, Model All-Amax Jr.
1924
Richard's Radios of Omaha

American-Bosch, Model #58
1930
Jay Daveler

American-Bosch, Model #355
1933
David & Julia Bart

American-Bosch, Model #670T
1937
Michael Feldt

Amsco Products, Melco Supreme
1924
Jim Berg

Apex, Model #8B
1931
Jim Berg

ARC, Model #515 New Zealand
1954
Ed & Irene Ripley

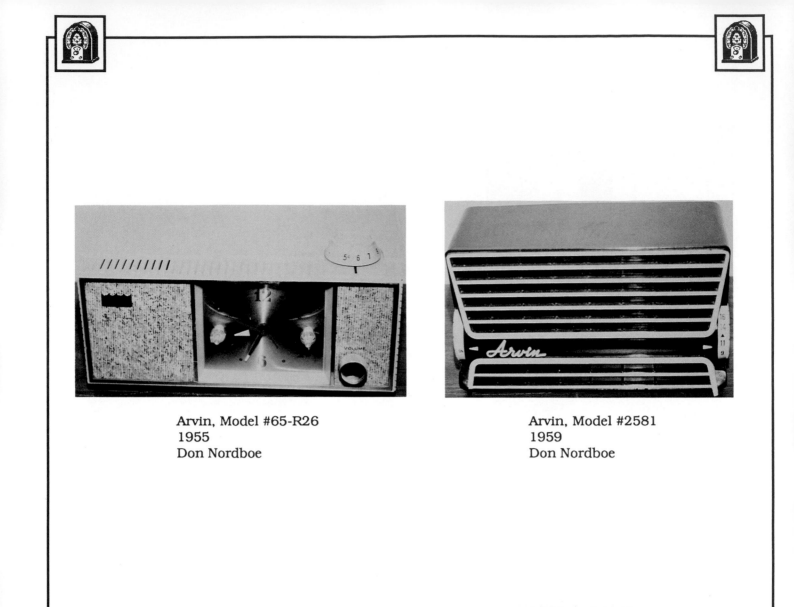

Arvin, Model #65-R26
1955
Don Nordboe

Arvin, Model #2581
1959
Don Nordboe

Arvin, Model #957T
1952
Don Nordboe

Arvin, Model #40 "Mighty Mite"
1938
Gary Wilson

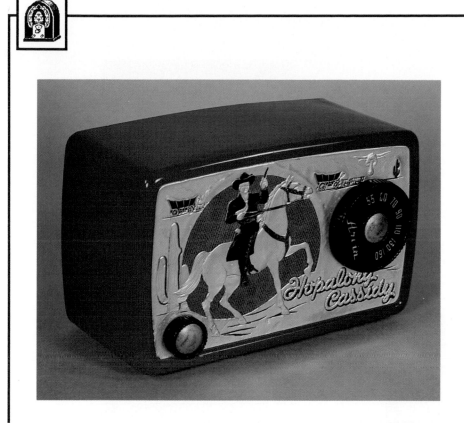

Arvin, Model #441-T
1950
Jay Daveler

Arvin, Model #956T
1956
Richard's Radios of Omaha

Arvin, Model #3561
1957
David & Julia Bart

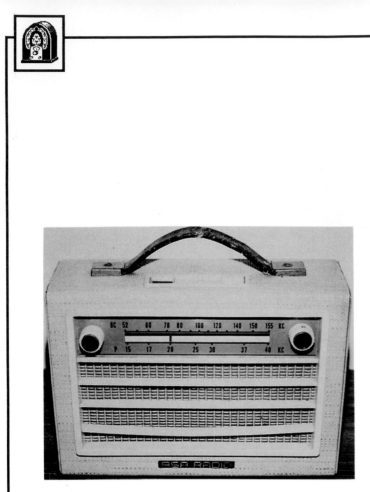

ASA, Model #TM155 Finland
1958
Don Nordboe

Arvin, Model #9577
1957
David Mednick

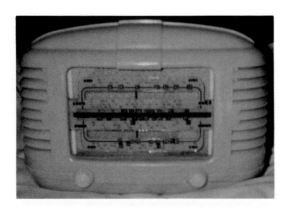

Astor, Australia
1930s
Steve Peterson

Astor, Model #4 Australia
1938
Don Nordboe

Atwater Kent, Model #5
1923
Larry & Nancy Anderson

Atwater Kent, Model #42
1928
David Kendall

Atwater Kent, Model #55
1929
Peter Oppenheim

Atwater Kent, Model #206
1934
Jay Daveler

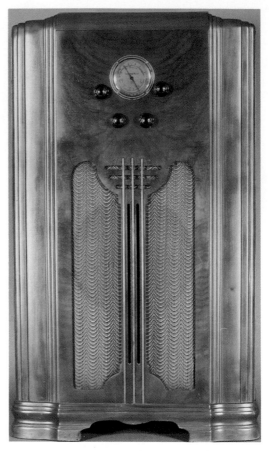

Atwater Kent, Model #317
1935
Jay Daveler

Atwater Kent, Model #387
1934
Michael Feldt

Atwater Kent, Model #856
1936
Doug Burskey

Atwater Kent, Model #318
1934
Spencer Doggett

Atwater Kent, Model #944
1934
Jay Daveler

Atwater Kent, Model #649
1935
Jim O'Neill

Automatiscope
1940
Mike Stambaugh

AWA, Radiola Australian
1940s
Richard's Radios of Omaha

Atwater Kent, Model # 509 Tunamatic
1935
Greg Farmer

Belknap, Model #7022
1940
Gerald Larsen

Belmont, Model #5D 128
1946
Mike Stambaugh

Belmont, Model #675
1934
Bud Shields

Belmont
1934
Michael Feldt

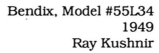

Bendix, Model #55L34
1949
Ray Kushnir

Bendix, Model #75P6U
1949
Ira & Debbie Grossman

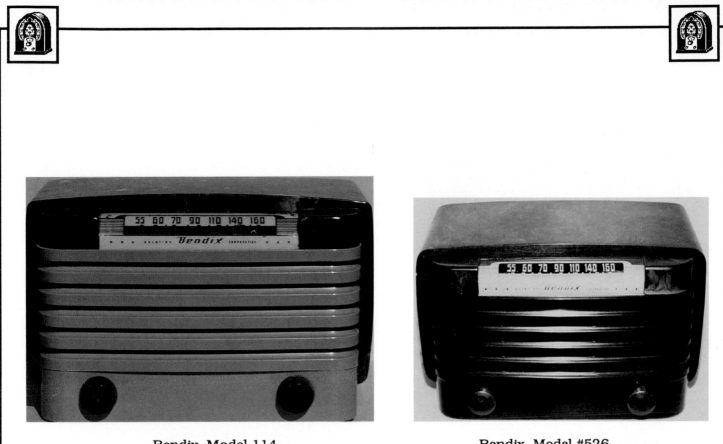

Bendix, Model 114
1948
Mike Stambaugh

Bendix, Model #526
1931
Richard's Radios of Omaha

Bendix, Model #526C
1946
Joe Ricci

Bendix, Model #753WX
1953
John Kendall

Blaupunkt, Model #GW646 Germany
1947
Max Kaplan

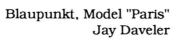
Blaupunkt, Model "Paris"
Jay Daveler

BRL, Japanese
1932
Max Kaplan

Brunswick, Model #1659
1940
Ed & Irene Ripley

Brunswick, Model #5-KR
1928
Spencer Doggett

Brunswick, Model #5-WO
1928
David & Julia Bart

Brush-McCoy, Bug Crystal Set
1927
Greg Farmer

Bulova Clock-Radio, Model #120
1957
Don Nordboe

C.A. Earl Radio Corp., Model #21
1929
Greg Nelson

Capehart, Model #1P55
1955
John E. Kendall

Carnival
1928
David & Julia Bart

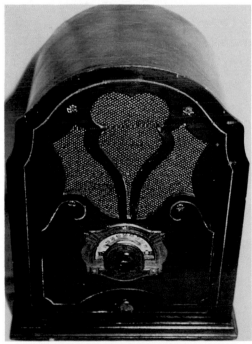

Century, Model #447
1931
Richard's Radios of Omaha

Chacophone, Model "Senior Two" England
1925
Max Kaplan

Channel Master, Model #6502
Richard's Radios of Omaha

Clarion, Model #AC-53
1929
Greg Nelson

Chun King, Model #ATR 210
1959
John E. Kendall

Clarion
1933
Mac's Old Time Radios

Clinton, Model #139
James Beaudet

Clinton, Model #216
1936
Jay Kinnard

Clinton, Model #T21
1950s
Jay Kinnard

Clinton, "Cub"
1937
Gerald Larsen

Clinton, LaSalle
1935
John Reinicke

Co-Radio, Coin Op
1940s
Gerald Larsen

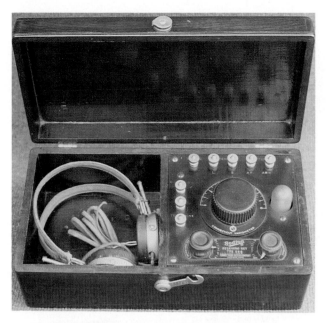

Connecticut Sodion, Model #DR-6
1923
Greg Farmer

Constellation, Model #10643
1960s
Don Nordboe

Continental, Model #1600
1952
Joe Ricci

Corona, Model #117
1936
Gerald Larsen

Coronado, Model #43-8351
1947
Spencer Doggett

Coronado, Model #RA-60-9917A
Richard's Radios of Omaha

Coronado, Model #A1
1937
Don Nordboe

Corsair, Model #347
1950
Ed & Irene Ripley

Crosley, Model #11-115-U
1951
Don Nordboe

Crosley, "Vanity"
1938
Don Nordboe

Crosley, Model #02CB
1942
Don Nordboe

Crosley, Model #02CA
1942
Don Nordboe

Crosley, Model #3R3 Trirdyn Special
1924
Jim Berg

Crosley, Model #9-103
1949
Mike Stambaugh

Crosley, Model #7M3
1935
Ed & irene Ripley

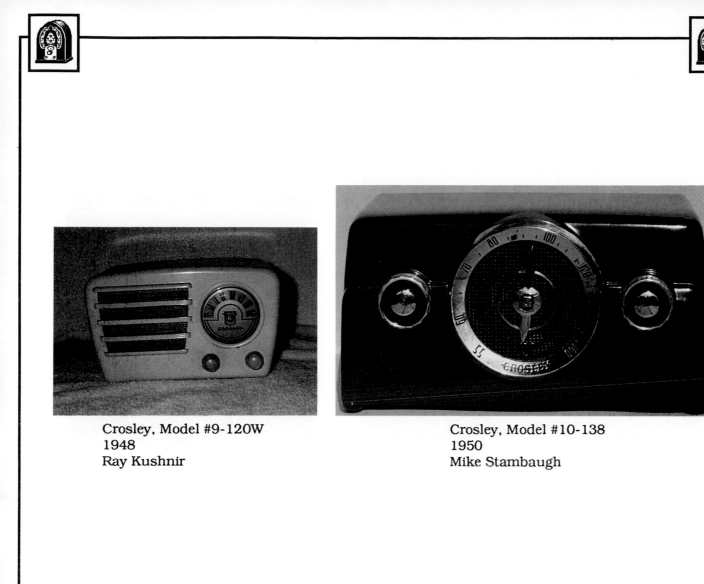

Crosley, Model #9-120W
1948
Ray Kushnir

Crosley, Model #10-138
1950
Mike Stambaugh

Crosley, Model #11-101-U
1951
Jay Daveler

Crosley, Model #11-111-U
Greg Farmer

Crosley, Model #11-118-U
1951
Jay Daveler

Crosley, Model #25-AW
Don Nordboe

Crosley, Model #52
1925
Jay Daveler

Crosley, Model #56-TP-L
1948
John Embry

Crosley, Model 56TX
1946
Mike Stambaugh

Crosley, Model #58 "Buddy Boy"
1931
Gary Hill

Crosley, Model #57TL
1948
David & Julia Bart

Crosley, Model #127
1931
John Embry

Crosley, Model #124
1931
Gary Wilson

Left - Crosley, Model #140 (Closed)
1931
David & Julia Bart

Bottom - Crosley, Model #140 (Open)
1931
David & Julia Bart

Crosley, Model #176 "Travette"
1933
Gary Wilson

Crosley, Model #178
1934
Michael Feldt

Crosley, Model #515
1935
Richard's Radios of Omaha

Crosley, Model #546
1936
Michael Feldt

Crosley, Model #555
1935
Jay Daveler

Crosley, Model #828-N
1939
Jay Daveler

Crosley, Model #729C
Ray Kushnir

Crosley, Model #ACE-3-B
1923
Jim Berg

Crosley, Model #C-529B
1939
Spencer Doggett

Crosley, Model #E-75TN
Don Nordboe

Crosley, Model #E-10BE
1953
Gerald Larsen

Crosley, Model #F25-BE
1952
Don Nordboe

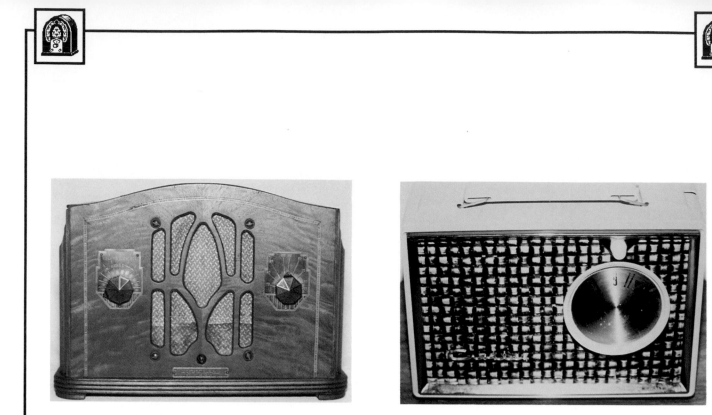

Crosley, "Companion"
1933
Bob & Margie Amos

Crosley, Model #P60-PK
1960
Don Nordboe

Crosley, Model #JC-6WEN
1956
Don Nordboe

Crosley, Model #JC-BWE
1957
Don Nordboe

Crosley, "Musical Chef"
1959
Jay Davcler

Cyarts, Model #B
1946
Bud Sheilds

Dahlberg, Model #430-DI Pillow Speaker Coin Op
1955
Bud Shields

Day Fan, Model #5091
1929
Jay Daveler

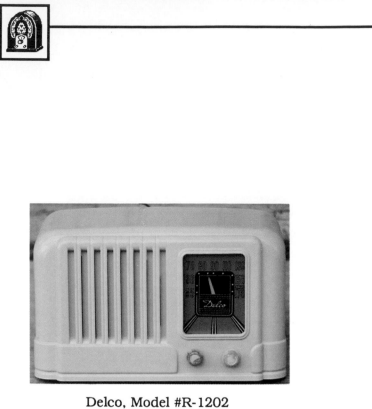

Delco, Model #R-1202
1941
Spencer Doggett

Decor, Bar-Adio
Jay Daveler

Deforest, Model #T-200
1920
David & Julia Bart

Delco, Model #R-1127
1938
Doug Burskey

Delco, Model #R1233
1940s
Mike Koste

Delco, Model #R1236
1947
Gerald Larsen

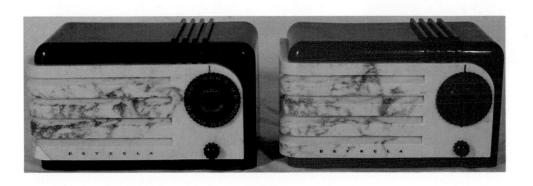

Detrola, Model #219
1938
Gary Wilson

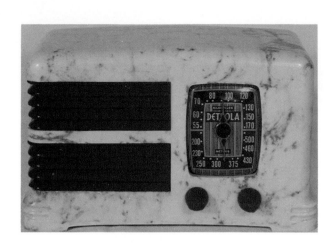

Detrola, Model #274
1939
Gary Wilson

Detrola, Model #280
1939
Gary Wilson

Dewald, Model #A-501
1938
Gary Wilson

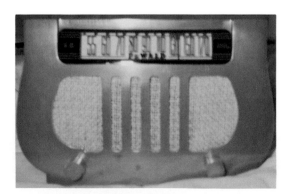

Dewald, Model #A-501
1930s
Steve Peterson

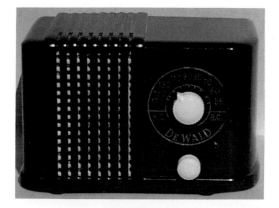

Dewald, Model #B-401
1948
Mike Stambaugh

Dewald, Model #B-512
Greg Farmer

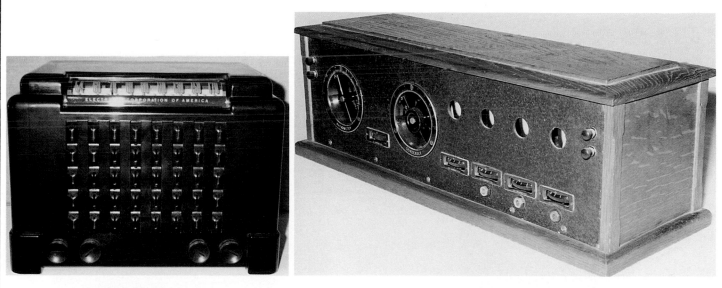

ECA, Model #108
1946
Bud Shields

Eisemann Magneto Corp.
1924
Max Kaplan

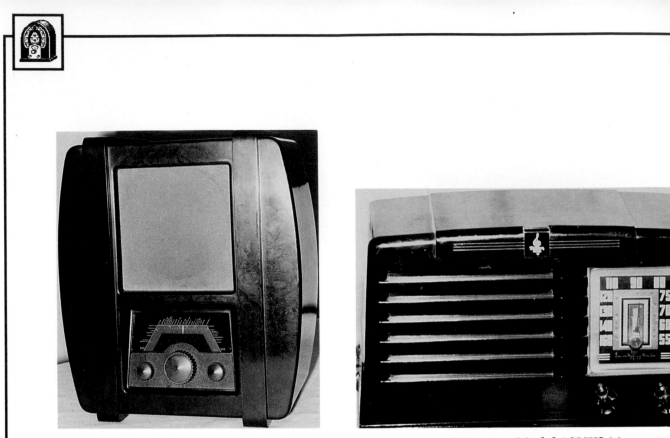

Ekco, Model #AD37 England
1930s
Max Kaplan

Emerson, Model #9LW344
1939
Gerald Larsen

Emerson, Model #23
1934
Jim Berg

Emerson, Model #38
1934
Richard's Radios of Omaha

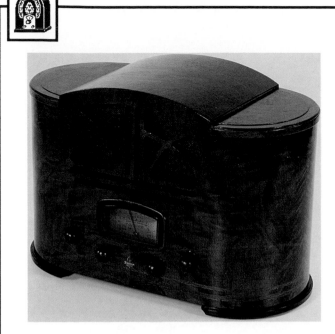

Emerson, Model #107
1935
Jay Daveler

Emerson, Model #149
1930s
Jim O'Neill

Emerson, Model #400 "Aristocrat"
1940
David Mednick

Emerson, Model #440
1941
Jay Daveler

Emerson, Model #522
1946
John Kendall

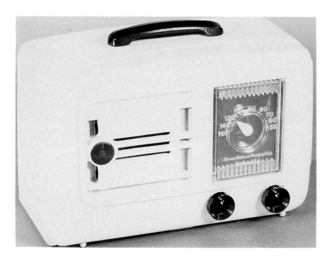

Emerson, Model #527
1947
Jay Daveler

Emerson, Model #535
1947
Gerald Larsen

Emerson, Model #547-A
Greg Farmer

Emerson, Model #564
1940
Ray Kushnir

Emerson, Model #572
1949
Gary Wilson

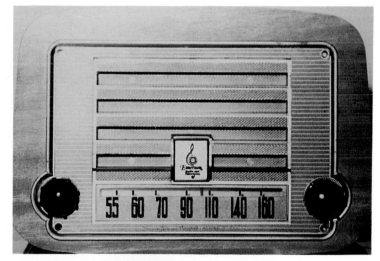

Emerson, Model #578-A
1946
Gerald Larsen

Emerson, Model #581-A
1949
Ira & Debbie Grossman

Emerson, Model #642
1950
Robert Rouette

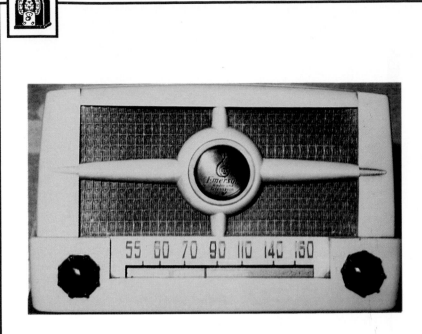

Emerson, Model #587-A
1950
Gerald Larsen

Emerson, Model #652
1952
Ira & Debbie Grossman

Emerson, Model #652-B
1950
Bob & Margie Amos

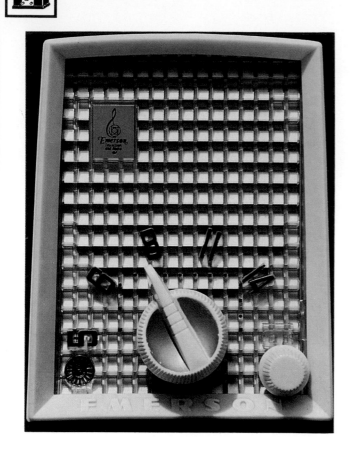

Emerson, Model #706-B
1952
Ira & Debbie Grossman

Emerson, Model #713
1953
Don Nordboe

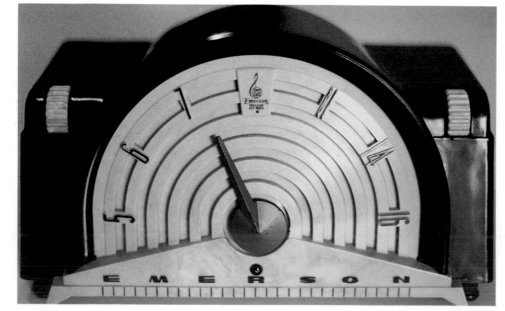

Emerson, Model #744-B
1954
David Mednick

Emerson, Model #744-B
1954
Mike Stambaugh

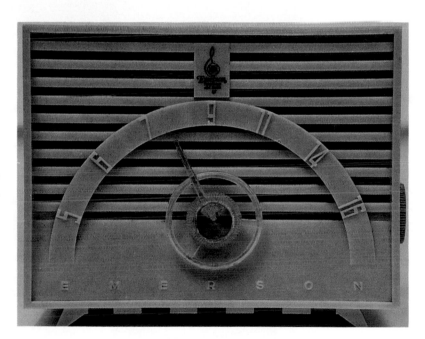

Emerson, Model #811
1955
Ira & Debbie Grossman

Emerson, Model #838
1956
David Mednick

Emerson, Model #888 "Explorer"
1959
David Mednick

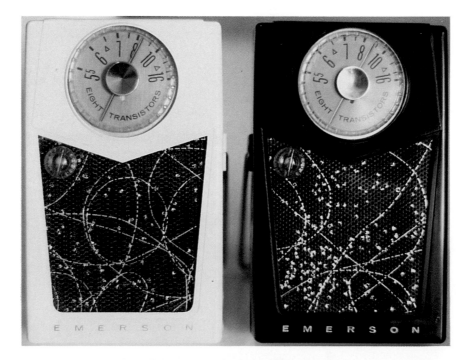

Emerson, Model #888 "Atlas"
1960
David Mednick

Emerson, Model #888 "Pioneer"
1958
David Mednick

Emerson, Model #888
"Vanguard"
1958
David Mednick

Emerson, Model #AX-211
1938
Greg Farmer

Emerson, Model #AX-243
1939
Ed & Irene Ripley

Emerson, Model #BA-199
1938
David & Julia Bart

Emerson, Model #BF-191
1938
Gerald Larsen

Emerson, Model #BM-258 "Big Miracle"
1937
Gary Wilson

Emerson, Model #BW-231
1938
Gary Wilson

Emerson, Model #CF-255
1938
Gerald Larsen

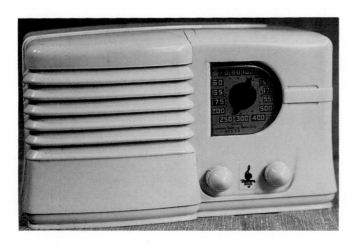

Emerson, Model #CH-246
1939
Ron Boucher

Emerson, Model #CH-256 "Strad"
1939
David & Julia Bart

Emerson, Model #CS-270
1939
Gerald Larsen

Emerson, Model #DB-315
1939
Gerald Larsen

Emerson, Model #DR-352
1941
Doug Burskey

Emerson, Model #DS
1939
Gerald Larsen

Emerson, Model #DY
1937
Gerald Larsen

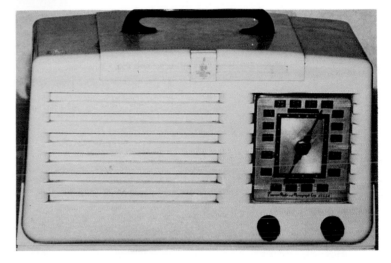

Emerson, Model #EC-301
1940
Peter Oppenheim

Emerson, Model #FU-428
1942
Doug Burskey

Emerson, Model #EC-853
1930s
John Kendall

Emerson, Model #TSI
1930s
Jay Daveler

Emerson "Burwood"
1931
Gary Hill

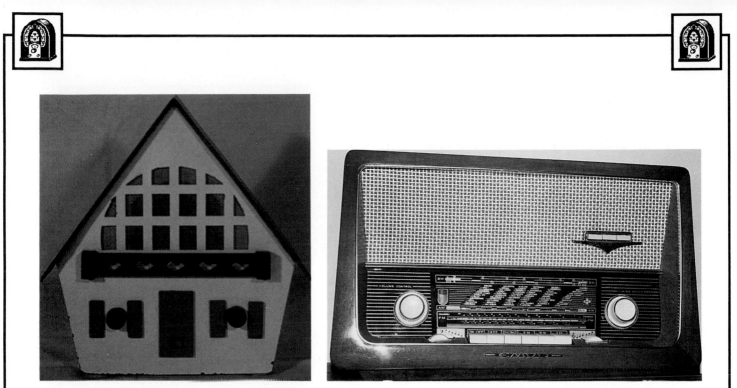

Empire Designing Corp., Model #55 Empress
1946
Gary Wilson

Emud, Model #60 Senior, West Germany
1959
David Kendall

Etherphone, Model #IV England
1926
Max Kaplan

Eveready, Model #3
1928
Arthur Kreitner

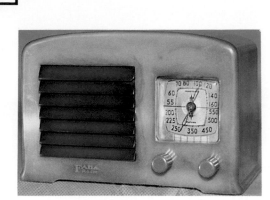

Fada, Model #53X Catalin
1938
John Kendall

Fada, Model #652 Catalin
1946
Davis Electronics

Fada, Model #652
1946
Davis Electronics

Fada, Model #659
1946
Gary Wilson

Fada, Model #740
1947
Gerald Larsen

Fada, Model #711
1947
Joe Ricci

Fada, Model #790
1948
David & Julia Bart

Fada, Model #845
1948
Gary Wilson

Fada, Model #845
1949
John Allport

Fada, Model #1000
1945
Ira & Debbie Grossman

Fairbanks-Morse, Model #8A
1936
Michael Feldt

Fairbanks-Morse, Model #5312
1934
Jay Daveler

Farnsworth, Model #AT
1939
Gerald Larsen

Farnsworth, Model #GT051
1948
Jay Daveler

Federal, Model "Junior" Crystal Set
1922
Greg Farmer

Federal, Model #1040TB
1946
Jay Daveler

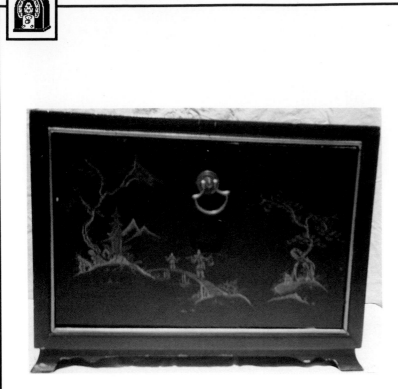

Federal, Model #1050T
1946
Jay Kinnard

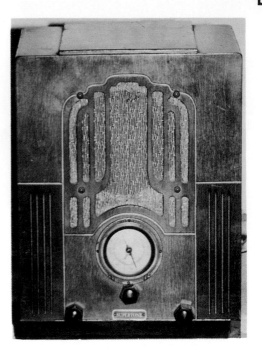

Federal, Model #A-42
1934
Gerald Larsen

Federal, Model #B20
Ross Mason

Firestone, Model #5-7798
1940s
Mike Koste

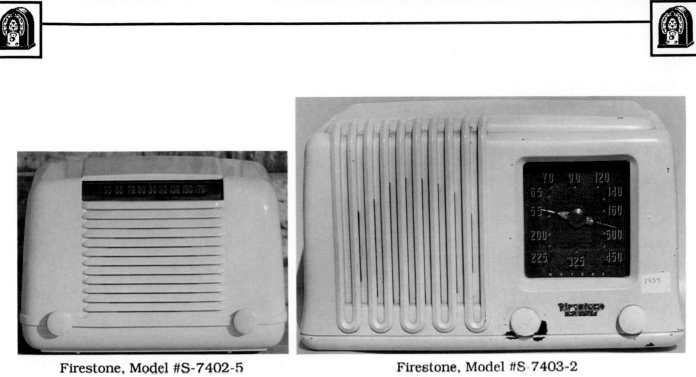

Firestone, Model #S-7402-5
1940
Spencer Doggett

Firestone, Model #S-7403-2
1939
Mike Stambaugh

Firestone, Model #S-7403-8
1939
Don Nordboe

Firestone
1940s
Mike Koste

Fleweling
1924
Richard's Radios of Omaha

Freed-Eisemann
1933
Gerald Larsen

Freed-Eisemann
1930s
Jay Daveler

Freed-Eisemann, Model "Sky Chief"
1930s
Jay Daveler

Garod, Model #5-A-1
1947
Mike Stambaugh

Garod, Model #SA2
1946
David & Julia Bart

Garod, Model #6AV1
"Commander"
1947
David Mednick

Garod, Model #6AV1 "Commander"
1945
Jim O'Neill

Garod, Model #6BU-1A
1947
Spencer Doggett

Geloso, Model #0.307 Italy
1958
Max Kaplan

General Electric, Model #60
1948
Bob & Margie Amos

General Electric, Model #221
1946
Gerald Larsen

General Electric, Model #405
1950
Gerald Larsen

General Electric, Model #548
1953
Joe Ricci

General Electric, Model #625
1950
Don Nordboe

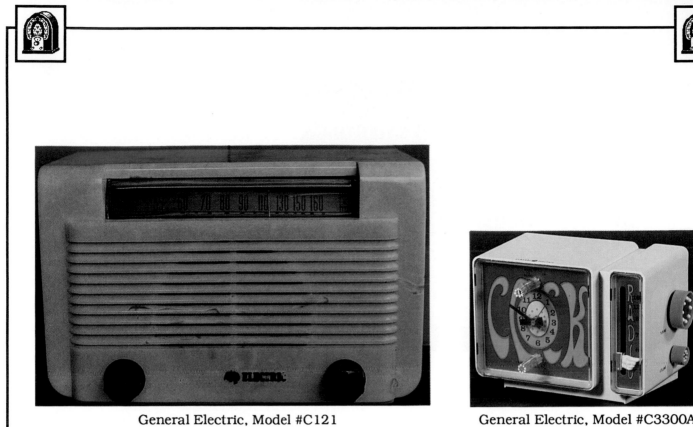

General Electric, Model #C121
1947
Robert Rouette

General Electric, Model #C3300A
Jay Daveler

General Electric, Model #C2418A
Jay Daveler

General Electric, Model #F74
1939
Arthur Kreitner

General Electric, Model #E50
1936
Gerald Larsen

General Electric, Model #G50
1937
Jay Daveler

General Electric, Model #GD60
1938
Jay Daveler

General Electric, Model #H116
1939
Spencer Doggett

General Electric
1935
Spencer Doggett

General Electric
1939
Spencer Doggett

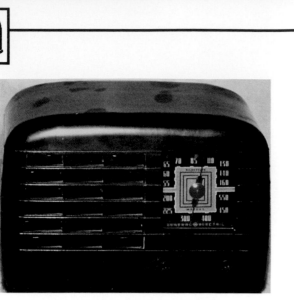

General Electric, Model #J644
1941
Gerald Larsen

General Electric, Model #K53
1933
Gerald Larsen

General Electric, Model #L50
1932
Greg Farmer

General Electric, Model #K64
1932
Gerald Larsen

General Electric, Model #J100
1932
Ken Deecken

General Electric, Model #L633
1942
Don Nordboe

General Electric, Model #L660
1942
Jay Daveler

General Electric, Model #M51
Michael Feldt

General Electric, Model #T110A
1950s
Bob & Margie Amos

General Electric "Sportmate"
1960s
Don Nordboe

General Motors
1931
Greg Farmer

General Television, Model #2A5
1938
Gerald Larsen

General Television, Model #534
1938
Gerald Larsen

General Television, Model #D-915
1939
Gerald Larsen

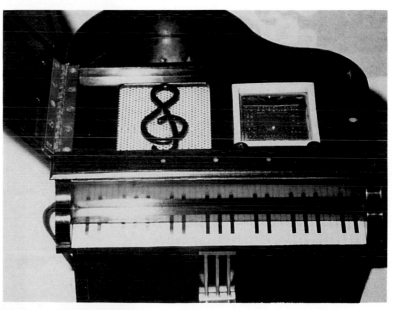

General Television
1940s
Bob & Margie Amos

General Television, "Victory"
1934
Gerald Larsen

Globe Radio Co.
1937
Gerald Larsen

Globetrotter
Greg Farmer

Gloritone, Model #26
1931
David & Julia Bart

Gloritone, Model #26PX
1930s
Jay Daveler

Grebe, Model #MI-2
1931
Greg Farmer

Grebe, Model "Synchrophase Seven"
1927
Mac's Old Time Radios

Grunow, Model #500
1933
Max Maldonado

Grunow, Model #569
1937
Gerald Larsen

Grunow, Model #1297
1936
Jay Daveler

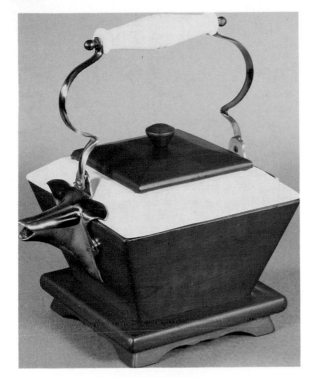

Guild "Teapot"
Jay Daveler

Guild "Hurdy Gurdy"
1950s
Gerald Larsen

Guild "Treasure Chest"
1950s
Gerald Larsen

Guild "Country Belle"
1950s
Gerald Larsen

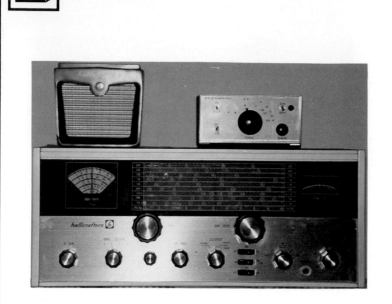

Hallicrafters, Model #SX-133
1972
David Lyons

Halton, Model #64-1
1945
Gerald Larsen

Herald
1928
David & Julia Bart

Howard, Model #118
1937
Doug Burskey

Hoffman, Model #BP706
1959
David Mednick

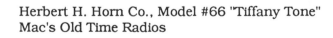

Herbert H. Horn Co., Model #66 "Tiffany Tone"
Mac's Old Time Radios

Howard, Model #225
1938
Doug Burskey

Howard, Model #901-A
1946
Gerald Larsen

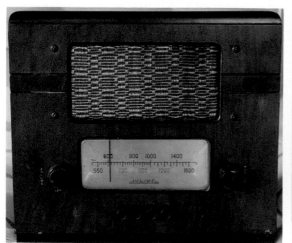

Howard, Model #300
1940
Doug Burskey

Howard, Model #780
1941
Doug Burskey

Howard, Model #906
1946
Doug Burskey

Howard, Model #906-S
1948
Gerald Larsen

Howard, Model #E-259
1936
Gerald Larsen

Imperial, Model #5 Germany
1933
Max Kaplan

Imperial, Model #500
1934
Gerald Larsen

Imperial
1932
David & Julia Bart

Imperial
1940s
Mike Stambaugh

JAX Cathedral
1932
Mac's Old Time Radios

Jewel
1950s
Ira & Debbie Grossman

Jewel, Model #955
1950
Ray Kushnir

Kadette, Model #35
1936
Gary Wilson

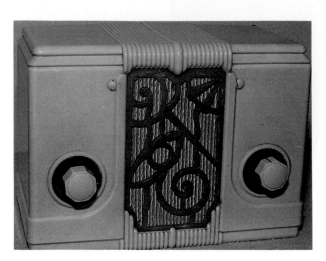

Kadette, Model #43 "Jewel"
1935
Greg Farmer

Kadette, Model #40 "Jewel"
1935
Gary Wilson

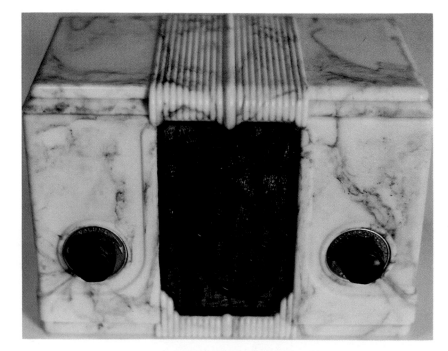

Kadette, Model #48 "Jewel"
1938
David Mednick

Kadette, Model #120
Greg Farmer

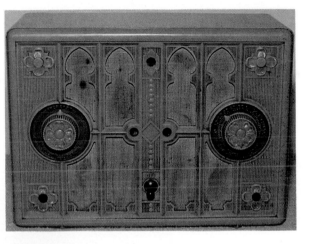

Kadette
1932
Greg Farmer

Kadette, Model #H
1931
David Mednick

Kadette, Model #K-25 "Clockette"
1938
David Mednick

Kadette, "Classic"
1936
Greg Farmer

Kadette, Model #L "Classic"
1936
David Mednick

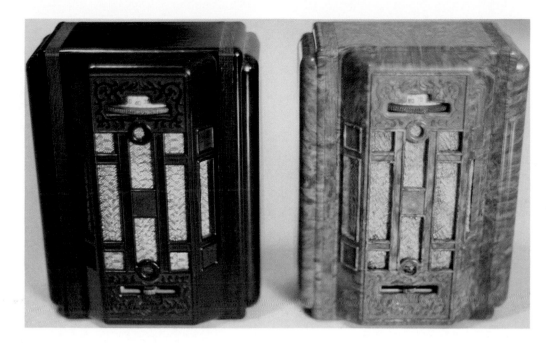

Kadette, "Junior"
1932
David Mednick

Kennedy, Model #XV
1925
Ross Mason

Kennedy, Model #110
1923
Ross Mason

Kennedy
1932
Michael Feldt

Knight, Model #168
1937
Don Nordboe

Knight, Model #5000
1946
Gerald Larsen

Knight, Model #A9700
1937
Gerald Larsen

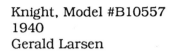

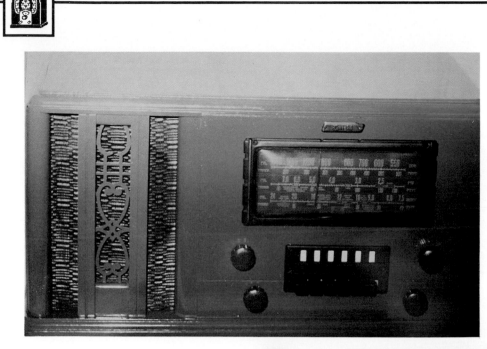

Knight, Model #B10557
1940
Gerald Larsen

Knight, Model #FA440
1938
Gerald Larsen

Knight, Model #G9511-13
1935
Gerald Larsen

Kowa, Model #KT-62
1962
Richard's Radios of Omaha

Kriesler, Australian
1934
Richard's Radios of Omaha

Lafayette, Model #JS178
1940
David & Julia Bart

Lafayette
1930s
Peter Oppenheim

Lincoln, Model #5A-110
1947
Gerald Larsen

Magnavox, Model #150
1927
Don Nordboe

Majestic, "Charlie McCarthy"
1938
Gerald Larsen

Majestic, Model #5-LA-5
1957
Mike Stambaugh

Majestic, Model #5-LA-80
1951
Ira & Debbie Grossman

Majestic, Model #7-P-420
1947
Doug Burskey

Majestic, Model #44
1933
Gerald Larsen

Majestic, Model #55
1933
Gerald Larsen

Majestic, Model #71
1928
Greg Nelson

Majestic, Model #77
1933
Gary Wilson

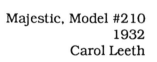

Majestic, Model #91
1929
Jim Berg

Majestic, Model #210
1932
Carol Leeth

Majestic, Model #300-A
1930s
Jay Daveler

Majestic, Model #651
1937
Gerald Larsen

Majestic, Model #921 "Melody Cruiser"
1946
Gary Wilson

Majestic
1934
Mac's Old Time Radios

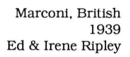

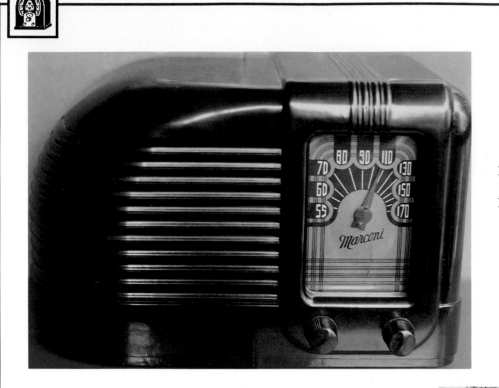

Marconi, Model #180
1939
Robert Rouette

Marconi, British
1939
Ed & Irene Ripley

Martin of California, "Planter Radio"
1948
Mac's old Time Radios

Meck, Model #RC-5C5-DL
1946
Gerald Larsen

Midland, Model #M6B
1946
Gerald Larsen

Midwest, Model #16-34
1934
Greg Farmer

Midwest, Model #17-39
1939
Jim Berg

Midwest, Model #18-36
1936
Spencer Doggett

Midwest, Model #62-B
1942
Richard's Radios of Omaha

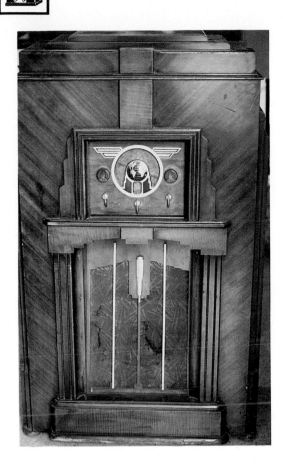

Midwest, Model #D-16
1935
Greg Farmer

Midwest, Model #T-6
1942
Greg Farmer

Midwest, Miraco Ultra 5
1925
Jim O'Neill

Midwest, "Royale"
1936
Greg Farmer

Mitchell, Model #1260
1940
Don Nordboe

Morrow, Model #CM-1
1950s
Gerald Larsen

Motorola, Model #5H-13
1952
Gerald Larsen

Motorola, Model #5T
1937
Doug Burskey

Motorola, Model #6X31
1957
John Kendall

Motorola, Model #50X1
Gerald Larsen

Motorola, Model #52CW2
1953
Bud Shields

Motorola, Model #53LC1
1954
Don Nordboe

Motorola, Model #55X11-A
1946
Ira & Debbie Grossman

Motorola, Model #56R3
1956
Gerald Larsen

Motorola, Model #58G1
1949
John Kendall

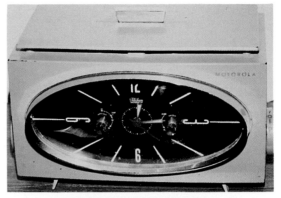

Motorola, Model #63C
1956
Don Nordboe

Motorola, Model #58L11
1948
Ira & Debbie Grossman

Motorola, Model #65X11A
1940s
Mark Byrd

Motorola, Model #68L11
1949
Doug Burskey

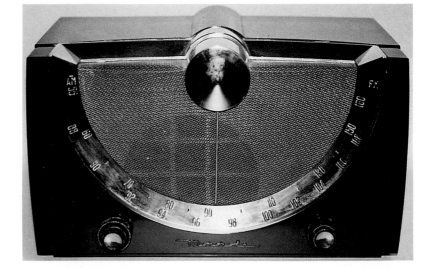

Motorola, Model #79XM21
1950
Larry Anderson

National, Model #SW-54
1951
Jay Daveler

Nora, Model #GW68 Germany
1938
Max Kaplan

Northland
1926
Greg Farmer

Olympic, Model #6-502
1946
Gerald Larsen

Olympic, Model #7-421W
1949
Mike Stambaugh

Olympic
1947
Gerald Larsen

Ozarka, Model #91-AC "Viking"
1932
Gerald Larsen

Packard-Bell, Model #5A5
1959
Ira & Debbie Grossman

Packard-Bell, Model #5DA
1947
Don Nordboe

Packard-Bell, Model #532
1954
Bob & Margie Amos

Paulson, Model #5-C
1935
Gerald Larsen

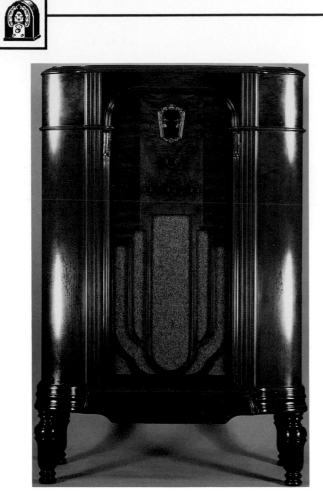

Philco, Model #14
1933
Jay Daveler

Philco, Model #20
1930
Gary Wilson

Philco, Model #37-93
1937
Jim O'Neill

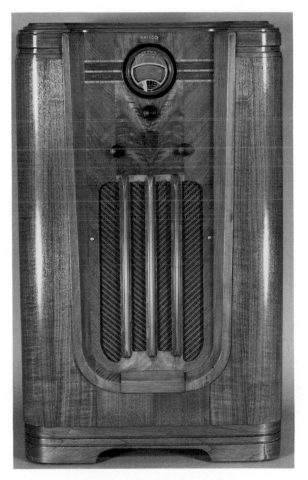

Philco, Model #37-630
1937
Jay Daveler

Philco, Model #37-650
1937
John Allport

Philco, Model #38-12
1938
Doug Burskey

Philco, Model #38-62
1938
David & Julia Bart

Philco, Model #38-610
1938
Greg Nelson

Philco, Model #39-6
1939
David & Julia Bart

Philco, Model #40-115
1940
Jay Daveler

Philco, Model #40-120
1940
Gerald Larsen

Philco, Model #40-135
1940
Gerald Larsen

Philco, Model #40-140
1940
Jay Daveler

Philco, Model #41-220
1941
Gerald Larsen

Philco, Model #40-180
1940
Spencer Doggett

Philco, Model #41-226
1941
John Allport

Philco, Model #41-230
1941
Mark Byrd

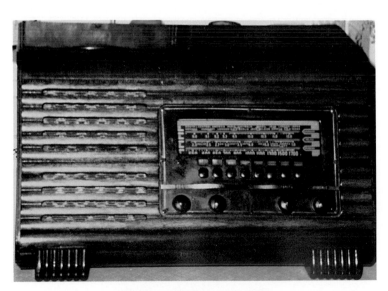

Philco, Model #41-250
1941
Gerald Larsen

Philco, Model #41-608
1941
Mark Byrd

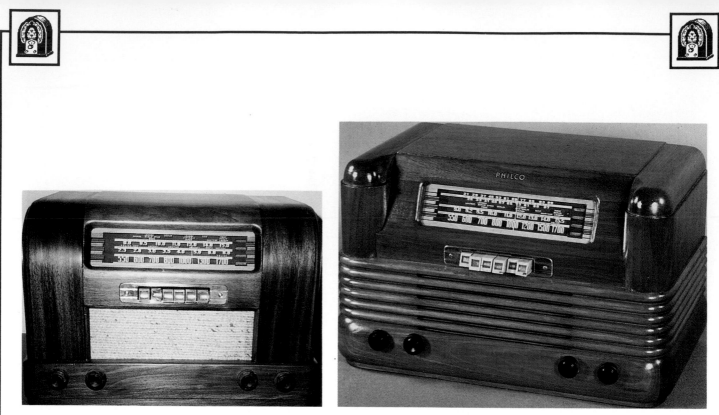

Philco, Model #42-345
1942
John Embry

Philco, Model #42-350
1942
Jay Daveler

Philco, Model #44
1933
David Kendall

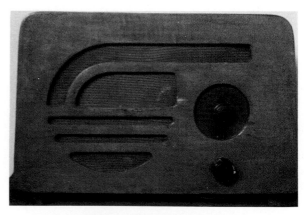

Philco, Model #46-431
1946
Peter Oppenheim

Philco, Model #47-205
1947
John Kendall

Philco, Model #48-464
1948
Gerald Larsen

Philco, Model #48-482
1948
Gerald Larsen

Philco, Model #49-503
1949
Mike Stambaugh

Philco, Model #50-621
1950
Bob & Margie Amos

Philco, Model #50-922
1950
Jay Kinnard

Philco, Model #51-538
1951
John Kendall

Philco, Model #52-642
1952
Jay Kinnard

Philco, Model #52-940
1952
Spencer Doggett

Philco, Model #52-942
1952
Gerald Larsen

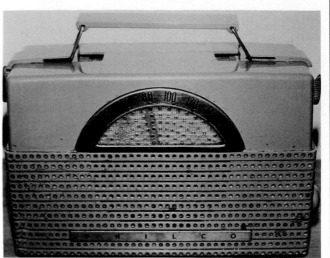

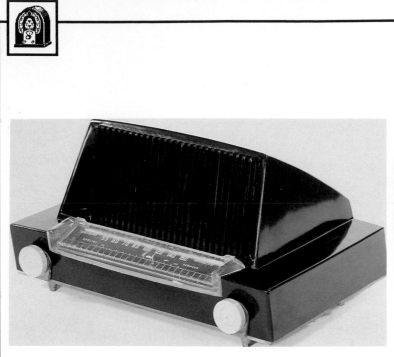

Philco, Model #53-566
1953
Jay Daveler

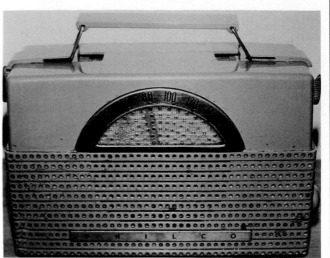

Philco, Model #53-652
1953
Don Nordboe

Philco, Model #60
1934
David & Julia Bart

Philco, Model #65
1929
Greg Nelson

Philco, Model # 70
1931
Jim Berg

Philco, Model #80 Jr.
1932
David Lyons

Philco, Model #95
1929
Jay Daveler

Philco, Model #112X
1931
Jay Daveler

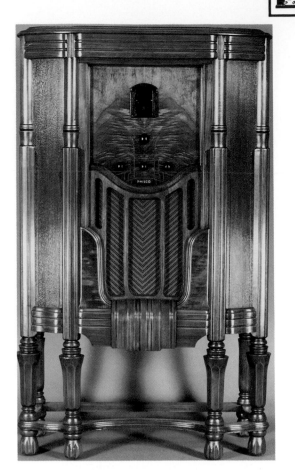

Philco, Model #118
1934
Jim Berg

Philco, Model #118
1935
Jay Daveler

Philco, Model #155
1940
Arthur Kreitner

Philco, Model #226C
1941
Jay Daveler

Philco, Model #620
1934
Bud Shields

Philco, Model #C582
1955
Gerald Larsen

Philco, Model #E976
1960
Gerald Larsen

Philco, Model #PT-2
1940
Gerald Larsen

Philco, Model #PT-6
1940
Gerald Larsen

Philco, Model #PT-25
1940
Gerald Larsen

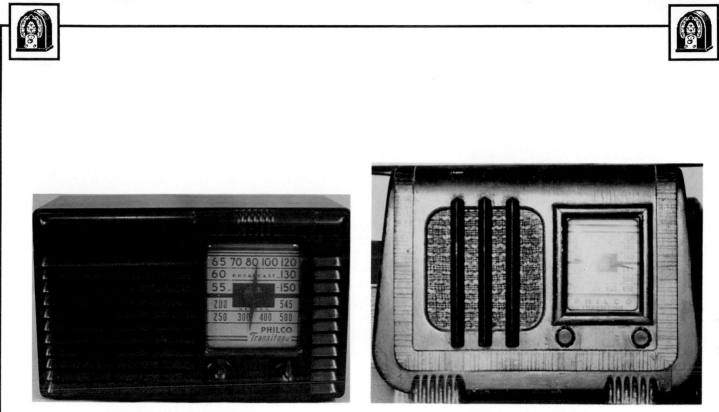

Philco, Model #PT-26
1940
David & Julia Bart

Philco, Model #PT-49
1941
Gerald Larsen

Philco, Model #PT-91
1941
Bob & Margie Amos

Philco, Model #T-1000-124
1959
Don Nordboe

Philmore, Model "Blackbird"
1933
Gary Wilson

Pilot, Model #T-1424
1940
Jay Daveler

Pilot, Model #P-235
1931
Carol Leeth

Pilot, Model #X-1252
Jay Daveler

Pilot, Model "All Wave"
1931
Ross Mason

PYE, Model #350/C England
1928
Max A. Kaplan

Radialva, Model #Super AS 56 France
1956
Don Nordboe

Radio Bell, Model #437 Belgium
1938
Max A. Kaplan

Radiola, Model #71-8
1947
Spencer Doggett

Radione, Model #R-3 Austria
1938
Max A. Kaplan

RCA, Model #5X
1936
Jay Daveler

RCA, Model #6K
1936
Spencer Doggett

RCA, Model #6T
1936
Jay Daveler

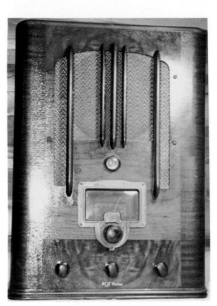

RCA, Model #6T5
1936
Jim Berg

RCA, Model #8X71
1949
Bob & Margie Amos

RCA, Model #8T "Victor"
1937
Jay Daveler

RCA, Model #9T "Victor"
1935
Jay Daveler

RCA, Model #8X53
1948
John Embry

RCA, Model #9TX31
1939
John E. Kendall

RCA, Model #16X4 "Victor"
1941
Jay Daveler

RCA, Model #19K
1940
Spencer Doggett

RCA, Model #Radiola 26
1924
Gary Hill

RCA, Model #40X53 "Siesta"
1939
Gerald Larsen

RCA, Model #29K2
1941
Jay Daveler

RCA, Model #44 Radiola
1929
Gary Wilson

RCA, Model #45X1
1940
Jay Daveler

RCA, Model #45X11
1940
Jay Daveler

RCA, Model #55X
1942
Jay Daveler

RCA, Model #95T5
1938
Jay Kinnard

RCA, Model #96T5
1939
Jay Kinnard

RCA, Model #110
1933
Richard's Radios of Omaha

RCA, Model #118
1935
Michael Feldt

RCA, Model #121
1933
Richard's Radios of Omaha

RCA, Model #222
1934
Ed & Irene Ripley

RCA, Model #B411
1954
Richard's Radios of Omaha

RCA, Model #T6-1
Michael Feldt

RCA, Model #T8-14
1935
Gerald Larsen

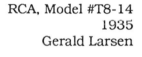

RCA, Model #Radiola III
1924
Jay Daveler

Realtone, Model #1088
"Comet"
1962
David Mednick

Regal, Model #C-527L
1952
Joe Ricci

Remler, Model #29
1929
Michael Feldt

Remler, Model #5400
1946
Ed & Irene Ripley

Remler, Model #5510
1946
Ed & Irene Ripley

Remler, Model #5565
1947
Ed & Irene Ripley

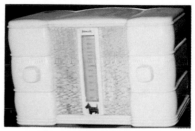

Remler, Model #46
1938
Ed & Irene Ripley

Scott Deluxe
1933
Ross Mason

Scott Super XII
1938
Greg Farmer

Sentinel, Model #238-V "Book Radio", (Open & Closed View)
1941
David & Julia Bart

Sentinel, Model #284-NI Catalin
1947

Setchell-Carlson, Model #58A-375
"Jet" 1940s
Bud Shields

Silver Marshall, Model #47A
James Beaudet

Silvertone, Model #6
1951
Don Nordboe

Silvertone, Model #132.878
1950
David & Julia Bart

Silvertone, Model #1470
1931
David & Julia Bart

Silvertone, Model #1660
1935
David & Julia Bart

Silvertone, Model #1906
1935
Spencer Doggett

Silvertone, Model #1964
Jay Daveler

Silvertone, Model #1965
1936
David & Julia Bart

Silvertone, Model #1967
Jay Daveler

Silvertone, Model #2005
1953
Spencer Doggett

Silvertone, Model #4501
1936
Ira & Debbie Grossman

Silvertone, Model #6110
1938
Ira & Debbie Grossman

Silvertone, Model #6421
1938
Bud Shields

Silvertone, Model #7054
1949
Bud Shields

Silvertone, Model #7072
1942
Jay Daveler

Silvertone, Model #7807
1939
Bud Shields

Silvertone, Model #8011
1949
Don Nordboe

Silvertone, Model #9001
1949
Mike Stambaugh

Silvertone, Superheterdyne
1930s
Peter Oppenheim

Silvertone
1930s
David & Julia Bart

Silvertone
1931
Mac's Old Time Radios

Simplex, Model #12198750
1933
David & Julia Bart

Sonora, Model #Comet KM
1941
Larry & Nancy Anderson

Sonora, Model #RBU-176
1946
Bud Shields

Sonora, Model #RCU-208
1946
Jay Daveler

Sonora, Model #WAV-243
1940
Jay Daveler

Sparton, Model #65
1934
Don Nordboe

Sparton, Model #136
1934
Jim Berg

Sparton, Model #410 "Jr."
1930
David & Julia Bart

Sparton, Model $506 "Bluebird"
1936
Gary Hill

Sparton, Model #1068
1937
Tom Ginocchio

Sparton, Model #557
1936
Greg Farmer

Spencer
1929
Richard's Radios of Omaha

St. Regis, Model #553
1935
Gerald Larsen

Sterling, Model "Deluxe"
1940s
David Mednick

Sterling, Model "Deluxe"
1940s
Greg Farmer

Stern Kit
1923
Max A. Kaplan

Stewart-Warner, Model #01-6D
1939
Mac's Old Time Radios

Stewart-Warner, Model #51T146
1947
Don Nordboe

Stewart-Warner, Model #61T26
1946
Bob & Margie Amos

Stewart-Warner, Model #305
1925
David & Julia Bart

Stewart-Warner, Model #325
1925
David & Julia Bart

Stewart-Warner, Model #1231-W
1934
Spencer Doggett

Stewart-Warner, Model #1451
1936
Jay Kinnard

Stewart-Warner, Model #9008A "Porta Bar"
1946
Don Nordboe

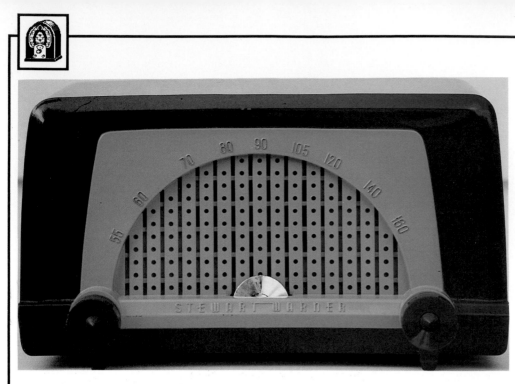

Stewart-Warner, Model #9161-A
1952
Ira & Debbie Grossman

Stewart-Warner, Model #9162
1952
Don Nordboe

Stewart-Warner, Model "Gulliver's Travels"
Greg Farmer

Stewart-Warner, Model #R-138A
1935
Jim Berg

Stewart-Warner, Model #R-1272-A
1934
Michael Feldt

Stewart-Warner
1934
David & Julia Bart

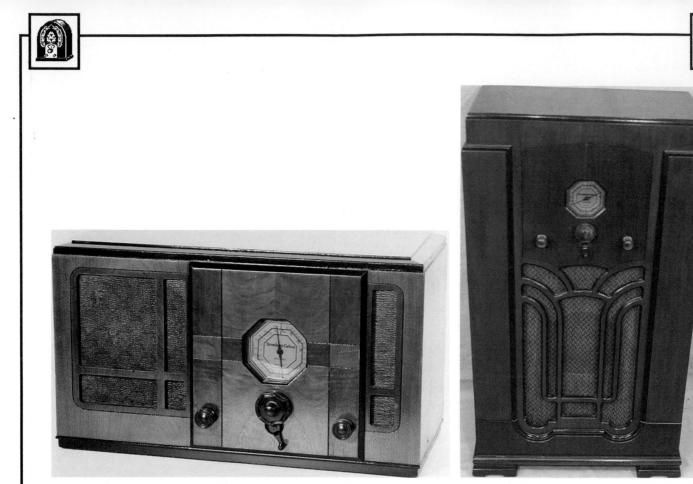

Stromberg-Carlson, Model #61-H
1935
Jay Daveler

Stromberg-Carlson, Model #58-L
1935
David & Julia Bart

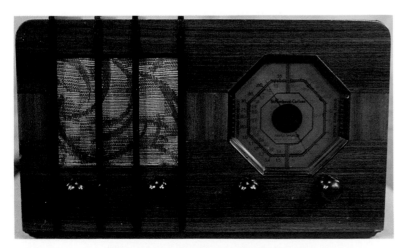

Stromberg-Carlson, Model #125-H
1936
David & Julia Bart

Stromberg-Carlson, Model #340-W
1937
Mark Bryd

Stromberg-Carlson, Model #652-A
Jay Daveler

Stromberg-Carlson, Model #1110-HW
1947
Jay Daveler

Stromberg-Carlson, Model #C-3
1955
Don Nordboe

Supertone
1935
Mac's Old Time Radios

Sylvania, Model #454
1954
Ira & Debbie Grossman

Sylvania, Model #596-3M
1953
Don Nordboe

Sylvania, Model #2207
1959
Don Nordboe

Sylvania, Model #U-235
Greg Farmer

Sylvania, Model "Thunderbird"
1957
David Mednick

Te-Ka-De, Germany
1955
Max A. Kaplan

Teletone, Model #195
1949
Ira & Debbie Grossman

Telefunken, Model #33WL Germany
1930
Max A. Kaplan

Telefunken, Germany
1950
Max A. Kaplan

Telefunken
1938
Max A. Kaplan

Telefunken
1960s
Don Nordboc

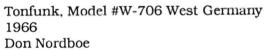

Tonfunk, Model #W-706 West Germany
1966
Don Nordboe

Toshiba, Model #7TH-425
1961
John E. Kendall

Toshiba, Model #9TL-3655
1965
Richard's Radios of Omaha

Tradio, Coin-Operated
1940s
Don Nordboe

Trav-ler
1935
Michael Feldt

Trav-ler, Model #T
1929
Jim Berg

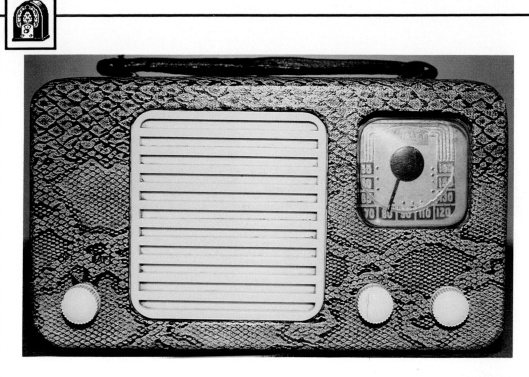

Trav-ler, Model #5028-A
1947
Ira & Debbie Grossman

Trophy, "Baseball Radio"
1941
Gary Hill

Troy
1938
Mac's Old Time Radios

Troy, "Blue Mirror"
1938
Mac's Old Time Radios

Truetone, Model #D-941
1937
Don Nordboe

Truetone, Model #D-2815
1948
Jay Daveler

U.S. Radio & Television, Model #41
1929
Greg Nelson

Waverly
1933
Mac's Old Time Radios

Western Electric, Model #7A
1925
Dennis Osborne

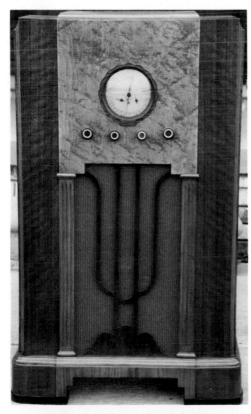

Wells-Gardner, Model #7D
1935
Spencer Doggett

Wega, Model "Fox II" Germany
1940s
Max A. Kaplan

Westingale
1926
Jim Berg

Westinghouse, Model #H104
1946
Jay Daveler

Westinghouse, Model #H157
1948
Bud Shields

Westinghouse, Model #H1G1 "Rainbow"
1948
Bud Shields

Westinghouse, Model #H204
1948
John E. Kendall

Westinghouse, Model #H211
1949
Don Nordboe

Westinghouse, Model #H563T6
1956
Jay Kinnard

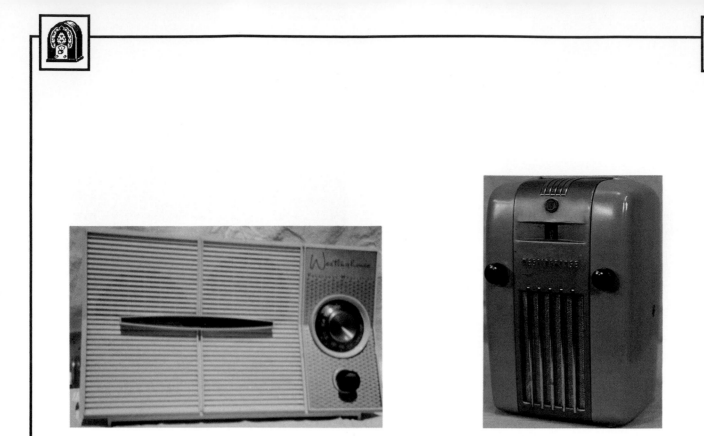

Westinghouse, Model #H716T5
1960
Jay Kinnard

Westinghouse, Model #H1251
1946
David & Julia Bart

Westinghouse, Model #WR14
1931
Ken Deecken

Westinghouse, Model #WR100
1935
Ed & Irene Ripley

Westinghouse, Model #WR232E
1938
Jim O'Neill

Westinghouse, Model #WR272
1939
Jay Daveler

Westinghouse, Model #WR274
1940
Doug Burskey

Westinghouse, Model #WR312
1936
Jay Daveler

Windsor, Model #BL006-P
1957
Richard's Radios of Omaha

Wurlitzer
1929
Mac's Old Time Radios

Zenith, Model #4G903
1948
John Embry

Zenith, Model #4R
1923
Greg Farmer

Zenith, Model #4V31
1935
David & Julia Bart

Zenith, Model #5J255
1938
Don Nordboe

Zenith, Model #5R216
1937
Jay Daveler

Zenith, Model #5S220
1938
Jay Daveler

Zenith, Model #6D015
1946
Spencer Doggett

Zenith, Model #6D315
1938
Jay Daveler

Zenith, Model #6D615
1942
Mikc Kostc

Zenith, Model 6D628
1943
John E. Kendall

Zenith, Model #6G05
1950
Spencer Doggett

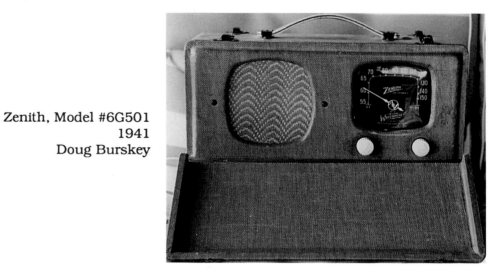

Zenith, Model #6G501
1941
Doug Burskey

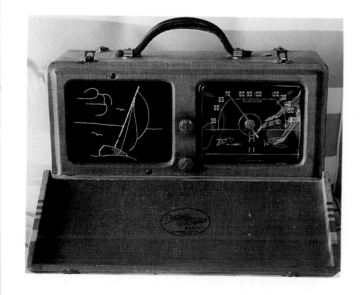

Zenith, Model # 6G601
1942
Doug Burskey

Zenith, Model #6G601M
1942
Jim Berg

Zenith, Model #6G801
1948
Doug Burskey

Zenith, Model #6S222
1938
David Kendall

Zenith, Model #6S232
1941
Jay Kinnard

Zenith, Model #6S239
1938
Steve Melvin

Zenith, Model #6S254
1938
Jay Daveler

Zenith, Model #6S511
1941
Jay Daveler

Zenith, Model #7E01
1940s
John E. Kendall

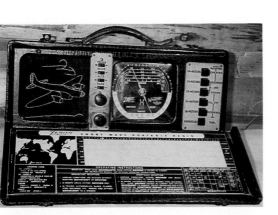

Zenith, Model #7G605 "Transoceanic"
1942
Jim Berg

Zenith, Model #7H820
1948
Jim Berg

Zenith, Model #7S261
1938
Jay Daveler

Zenith, Model #7S342 "Chairside"
1939
Don Nordboe

Zenith, Model #7S529
1941
David & Julia Bart

Zenith, Model #7S633
1942
Greg Nelson

Zenith, Model #8D015
1946
Ira & Debbie Grossman

Zenith, Model #8H034
1946
Jay Daveler

Zenith, Model #8S432
1940
Jay Daveler

Zenith, Model #8S463
1940
Steve Melvin

Zenith, Model #9S367
1938
Mark Bryd

Zenith, Model #10S155
1936
Jay Daveler

Zenith, Model #12S266
1938
Jim Berg

Zenith, Model #12U158
1937
Jim Berg

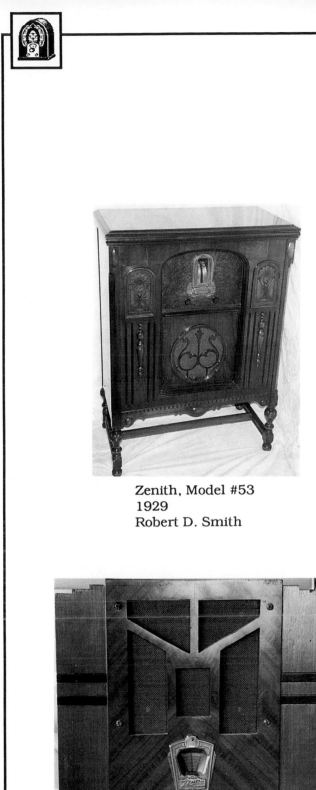

Zenith, Model #53
1929
Robert D. Smith

Zenith, Model #72X
1930
Jay Daveler

Zenith, Model #288
1934
Doug Burskey

Zenith, Model #400 "Royal"
1963
John E. Kendall

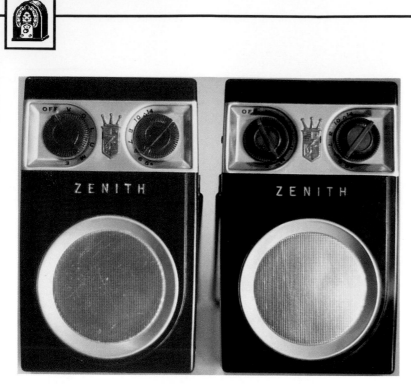

Zenith, Model #500 "Royal"
1955
David Mednick

Zenith, Model #500H "Royal"
1962
John E. Kendall

Zenith, Model #750 "Royal"
1959
Jay Daveler

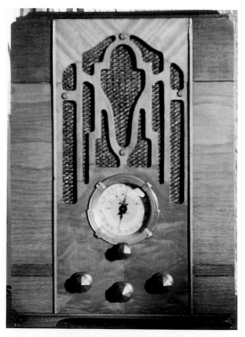

Zenith, Model #807
1933
Ted LaFleur

Zenith, Model #970
1934
Jay Daveler

Zenith, Model #A
1932
Ron Boucher

Zenith, Model #H-615
1951
Spencer Doggett

Zenith, Model #H500 "Transoceanic"
1951
David Lyons

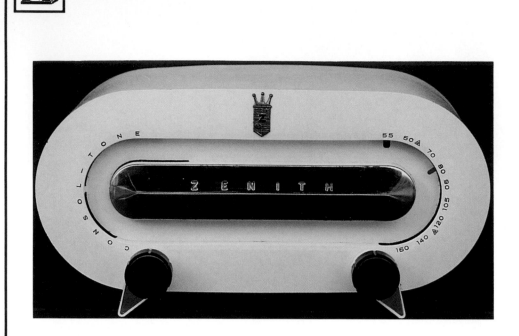

Zenith, Model #H511W
1951
Ira & Debbie Grossman

Zenith, Model #J616
1952
Don Nordboe

Zenith, Model #J733
1952
Don Nordboe

Zenith, Model #J402
1952
Don Nordboe

Zenith, Model #K412R
1953
Mike Stambaugh

Zenith, Model #520R
1954
Ira & Debbie Grossman

Zenith, Model #M505
1953
Don Nordboe

Zenith, Model #R511W
1955
Bob & Margie Amos

Zenith, Model #R615
1952
Jay Daveler

Zenith, Model #T505
1955
Ira & Debbie Grossman

Zenith, Model #T825
1956
Mikc Koste

Zenith, Model #Z733
1956
Spencer Doggett

Microphones
Jay Daveler

Gibbs Carbon Microphone
1923
David Kendall

General Electric Microphone, Model #55
1947
David Kendall

Champion Spark Plug Microphone
Jay Daveler

Ring Microphones
1920s
David & Julia Bart

Left - Electravolta, Model #731, Right-Shure, Model #55
David & Juiia Bart

RCA Ribbon Microphone, Model #BK11A
1948
David Kendall

Altec Microphone, Model "Salt & Pepper"
1949
David Kendall

Magnavox Speaker, Model #R-3
1921
David & Julia Bart

Atwater Kent Speaker, Model #H
1925
David & Julia Bart

Philco Speaker
Ira & Debbie Grossman

Left - Reichman Co. Speaker, Model #3 "Thorola
Sr., Right - Reichman Co. Speaker, Model "Thorola
Jr."
David & Julia Bart

Music Maker Speaker, Model #VI
1924
David & Julia Bart

Crosley Dynacone Speaker, Model #E
1927
David & Julia Bart

Radiola Loop Antenna,
Model #AG-814
Jay Daveler

Page 9
Reference Only
Page 10
$125-$175
Page 11
Top Left - $125-$250
Top Right - $175
Bottom Left - $175
Bottom Right - $190
Page 12
Top Left - $225
Top Right - $240
Bottom Left - $150
Bottom Right - $150
Page 13
$100-$200
Page 14
$150-$200
Page 15
$100-$200
Page 16
Top Left - $75-$125
Top Right - $100-$175
Bottom Left - $200-$300
Bottom Right - $150-$200
Page 17
$100-$200
Page 18
Top Left - $200
Top Right - $175
Bottom Left - $200
Bottom Right - $100-$200
Page 19
$75-$150
Page 20
$75-$150
Page 21
$50-$150
Page 22
$100-$200
Page 23
$50-$100
Page 24
$75-$150
Page 25
$50-$75
Page 26
$50-$100
Page 27
$75-$150
Page 28
$50-$100
Page 29
$35-$65
Page 30
Top - $25
Middle - $125

Bottom - $100
Page 31
Top - $110
Middle - $40
Bottom - $40
Page 32
Top - $40
Middle - $50
Bottom - $35
Page 33
Top - $50
Middle - $110
Bottom - $50
Page 34
Top - $65
Middle - $55
Bottom - $50
Page 35
Top Left - $60
Top Right - $50
Bottom Left - $35
Bottom Right - $55
Page 36
Top - $60
Middle - $110
Bottom - $75
Page 37
Top Left - $150
Top Right - $125
Bottom Left - $150
Bottom Right - $75
Page 38
Top Left - $200
Top Right - $175
Bottom Left - $225
Bottom Right - $100
Page 39
Top Left - $35
Top Right - $35
Bottom Left - $30
Bottom Right - $115
Page 40
Top - $550
Middle - $45
Bottom - $35
Page 41
Top Left - $50
Top Right - $55
Bottom Left - $1,500
Bottom Right - $75
Page 42
Top - $7,000
Middle - $100
Bottom - $175
Page 43
Top Left - $275
Top Right - $175

Bottom Left - $325
Bottom Right - $150
Page 44
Top - $150
Bottom - $325
Page 45
Top Left - $225
Top Right - $65
Bottom Left - $250
Bottom Right - $325
Page 46
Top - $45
Middle - $200
Bottom - $150
Page 47
Top - $100
Middle - $50
Bottom - $50
Page 48
Top Left - $300
Top Right - $500
Bottom Left - $575
Bottom Right - $70
Page 49
Top - $80
Middle - $60
Bottom - $100
Page 50
Top - $175
Middle - $125
Bottom - $175
Page 51
Top - $700
Middle - $50
Bottom - $150
Page 52
Top Left - $40
Top Right - $100
Bottom Left - $150
Bottom Right - $150
Page 53
Top Left - $40
Top Right - $150
Bottom Left - $45
Bottom Right - $225
Page 54
Top - $100
Middle - $100
Bottom - $40
Page 55
Top - $70
Middle - (Rare)
Bottom - $75
Page 56
Top Left - $300
Top Right - $80
Bottom Left - $35

Bottom Right - $50
Page 57
Top - $55
Middle - $30
Bottom - $125
Page 58
Top Left - $150
Top Right - $115
Bottom Left - $50
Bottom Right - $125
Page 59
Top Left - $100
Top Right - $125
Bottom Left - $50
Bottom Right - $175
Page 60
Top Left - $50
Top Right - $125
Bottom Left - $125
Bottom Right - $125
Page 61
Top Left - $110
Top Right - $60
Bottom Left - $150
Bottom Right - $50
Page 62
Top Left - $60
Top Right - $400
Bottom Left - $75
Bottom Right - $275
Page 63
Top - $250
Bottom - $325
Page 64
Top Left - $125
Top Right - $80
Bottom Left - $110
Bottom Right - $100
Page 65
Top Left - $140
Top Right - $200
Bottom Left - $90
Bottom Right - $150
Page 66
Top Left - $75
Top Right - $45
Bottom Left - $75
Bottom Right - $55
Page 67
Top Left - $110
Top Right - $45
Bottom Left - $50
Bottom Right - $100
Page 68
Top Left - $75
Top Right - $2,000
Bottom Left - $275

Bottom Right - $175
Page 69
Top Left - $50
Top Right - $200
Bottom Left - $400
Bottom Right - $75
Page 70
Top - $75
Middle - $35
Bottom - $500 each
Page 71
Top Left - $450
Top Right - $150
Bottom Left - $750
Bottom Right - $750
Page 72
Top Left - $80
Top Right - $750
Bottom Left - $65
Bottom Right - $150
Page 73
Top Left - $150
Top Right - $40
Bottom Left - $75
Bottom Right - $125
Page 74
Top - $135
Middle - $65
Bottom - $800
Page 75
Top Left - $60
Top Right - $55
Bottom Left - $50
Bottom Right - $30
Page 76
Top Left - $75
Top Right - $700
Bottom Left - $150
Bottom Right - $60
Page 77
Top - $50
Bottom - $50
Page 78
Top - $30
Middle - $50
Bottom - $50
Page 79
Top - $55
Middle - $80
Bottom - $250
Page 80
Top - $250
Middle - $50
Bottom - $75
Page 81
Top - $75 each
Bottom - $75 each

Page 82
Top - $75
Middle - $75 each
Bottom - $90
Page 83
Top Left - $125
Top Right - $125
Bottom Left - $45
Bottom Right - $1,200
Page 84
Top - $135
Middle - $90
Bottom - $95
Page 85
Top Left - $435
Top Right - $55
Bottom Left - $45
Bottom Right - $75
Page 86
Top - $55
Middle - $40
Bottom - $65
Page 87
Top Left - $175
Top Right - $50
Bottom Left - $200
Bottom Right - $800
Page 88
Top Left - $250
Top Right - $100
Bottom Left - $300
Bottom Right - $70
Page 89
Top Left - $2,500
Top Right - $650
Middle Left - $450
Middle Right - $450
Bottom Left - $60
Bottom Right - $50
Page 90
Top - $100
Middle - $250
Bottom - $225
Page 91
Top - $750
Middle - $125
Bottom - $150
Page 92
Top Left - $100
Top Right - $85
Bottom Left - $250
Bottom Right - $55
Page 93
Top Left - $250
Top Right - $80
Bottom Left - $150
Bottom Right - $75

Page 94
Top Left - $40
Top Right - $70
Bottom Left - $100
Bottom Right - $40
Page 95
Top - $150
Middle - $55
Bottom - $100
Page 96
Top - $100
Middle - $60
Bottom - $175
Page 97
Top - $1,500
Bottom - $700
Page 98
Top Left - $40
Top Right - $45
Bottom Left - $40
Bottom Right - $40
Page 99
Top - $45
Middle - $35
Bottom - $40
Page 100
Top Left - $75
Top Right - $70
Bottom Left - $75
Bottom Right - $70
Page 101
Top - $125
Middle - $75
Bottom - $80
Page 102
Top - $250
Middle - $130
Bottom - $125
Page 103
Top Left - $45
Top Right - $100
Middle - $150
Bottom Left - $275
Bottom Right - $300
Page 104
Top - $65
Middle - $70
Bottom - $150
Page 105
Top Left - $25
Top Right - $50
Bottom Left - $500
Bottom Right - $40
Page 106
Top - $350
Middle - $70
Bottom - $75

Page 107
Top Left - $150
Top Right - $450
Bottom Left - $450
Bottom Right - $250
Page 108
Top - $225
Bottom - $400
Page 109
Top Left - $175
Top Right - $175
Bottom Left - $60
Bottom Right - $250
Page 110
Top Left - $130
Top Right - $250
Bottom Left - $175
Bottom Right - $90
Page 111
Top - $125
Middle - $40
Bottom - $125
Page 112
Top - $125
Middle - $75
Bottom - $150
Page 113
Top Left - $50
Top Right - $30
Bottom Left - $50
Bottom Right - $50
Page 114
Top - $60
Middle - $65
Bottom - $300
Page 115
Top Left - $150
Top Right - $75
Bottom Left - $150
Bottom Right - $50
Page 116
Top - $250
Middle - $50
Bottom - $60
Page 117
Top Left - $100
Top Right - $250
Bottom Left - $150
Bottom Right - $250
Page 118
Top - $100
Middle - $300
Bottom - $250
Page 119
Top Left - $1,500
Top Right - $600
Bottom - $600

Page 120
Top - $250 each
Botttom Left - $400
Bottom Right - $800
Page 121
Top Left - $225
Top Right - $175
Bottom Left - $40
Bottom Right - $40
Page 122
Top - $60
Middle - $60
Bottom - $50
Page 123
Top Left - $30
Top Right - $250
Bottom Left - $100
Bottom Right - $75
Page 124
Top Left - $35
Top Right - $500
Bottom Left - $1,200
Bottom Right - $100
Page 125
Top - $60
Middle - $50
Bottom - $150
Page 126
Top - $200
Bottom Left - $225
Bottom Right - $175
Page 127
Top - $175
Middle - $600
Bottom - $900
Page 128
Top - $125
Middle - $500
Bottom - $250
Page 129
Top - $60
Middle - $200
Bottom - $125
Page 130
Top - $45
Middle - $40
Bottom - $350
Page 131
Top - $350
Middle - $375
Bottom - $125
Page 132
Top - $375
Middle - $75
Bottom - $175
Page 133
Top Left - $750

Top Right - $200
Bottom Left - $35
Bottom Right - $35
Page 134
Top Left - $100
Top Right - $70
Bottom Left - $30
Bottom Right - $35
Page 135
Top - $40
Middle - $40
Bottom - $30
Page 136
Top Left - $35
Top Right - $30
Bottom Left - $40
Bottom Right - $30
Page 137
Top - $65
Middle - $60
Bottom - $50
Page 138
Top - $200
Middle - $175
Bottom - $50
Page 139
Top - $75
Middle - $40
Bottom - $225
Page 140
Top Left - $35
Top Right - $55
Bottom Left - $30
Bottom Right - $50
Page 141
Top - $135
Bottom - $130
Page 142
Top - $130
Bottom - $160
Page 143
Top Left - $150
Top Right - $55
Bottom Left - $80
Bottom Right - $120
Page 144
Top Left - $50
Top Right - $50
Bottom Left - $55
Bottom Right - $55
Page 145
Top - $60
Middle - $125
Bottom - $40
Page 146
Top Left - $155
Top Right - $60

Bottom Left - $85
Bottom Right - $125
Page 147
Top Left - $65
Top Right - $85
Bottom Left - $185
Bottom Right - $65
Page 148
Top - $60
Middle - $40
Bottom - $50
Page 149
Top Left - $100
Top Right - $45
Bottom Left - $50
Bottom Right - $40
Page 150
Top - $35
Middle - $45
Bottom - $45
Page 151
Top Left - $60
Top Right - $35
Bottom Left - $160
Bottom Right - $125
Page 152
Top Left - $400
Top Right - $125
Bottom Left - $325
Bottom Right - $250
Page 153
Top Left - $325
Top Right - $200
Bottom Left - $90
Bottom Right - $125
Page 154
Top - $150
Middle - $30
Bottom - $30
Page 155
Top - $35
Middle - $45
Bottom - $35
Page 156
Top Left - $40
Top Right - $55
Bottom Left - $50
Bottom Right - $100
Page 157
Top Left - $175
Top Right - $75
Bottom Left - $350
Bottom Right - $75
Page 158
Top Left - $1,500
Top Right - $200
Bottom Left - $75

Bottom Right - $125
Page 159
Top Left - $40
Top Right - $75
Bottom Left - $125
Bottom Right - $130
Page 160
Top Left - $150
Top Right - $125
Bottom Left - $40
Bottom Right - $125
Page 161
Top Left - $40
Top Right - $200
Bottom Left - $65
Bottom Right - $60
Page 162
Top Left - $150
Top Right - $375
Bottom Left - $400
Bottom Right - $225
Page 163
Top - $110
Middle - $40
Bottom - $55
Page 164
Top Left - $60
Top Right - $75
Bottom Left - $65
Bottom Right - $225
Page 165
Top Left - $140
Top Right - $250
Bottom Left - $150
Bottom Right - $35
Page 166
Top - $115
Middle - $150
Bottom - $125
Page 167
Top - $50 each
Middle - $35
Bottom - $400
Page 168
Top Left - $125
Top Right - $150
Middle Left - $75
Middle Right - $75
Bottom Left - $350
Bottom Right - $700
Page 169
Top - $200
Bottom - $800
Page 170
Top - $500
Middle - $250
Bottom - $30

Page 171
Top Left - $60
Top Right - $275
Bottom Left - $165
Bottom Right - $125
Page 172
Top Left - $150
Top Right - $200
Bottom Left - $200
Bottom Right - $30
Page 173
Top - $75
Middle - $1,000
Bottom - $75
Page 174
Top Left - $50
Top Right - $110
Bottom Left - $100
Bottom Right - $40
Page 175
Top Left - $45
Top Right - $100
Bottom Left - $140
Bottom Right - $250
Page 176
Top Left - $200
Top Right - $1,400
Bottom Left - $45
Bottom Right - $70
Page 177
Top Left - $150
Top Right - $150
Bottom Left - $400
Bottom Right - $275
Page 178
Top - $3,000
Middle - $200
Bottom - $2,500
Page 179
Top - $100
Middle - $75
Bottom - $200
Page 180
Top Left - $200
Top Right - $150
Bottom Left - $275
Bottom Right - $30
Page 181
Top - $40
Middle - $165
Bottom - $115
Page 182
Top - $125
Middle - $60
Bottom - $325
Page 183
Top - $40

Middle - $40
Bottom - $800
Page 184
Top - $260
Middle - $175
Bottom - $175
Page 185
Top Left - $85
Top Right - $165
Bottom Left - $110
Bottom Right - $225
Page 186
Top - $700
Middle - $75
Bottom - $45
Page 187
Top Left - $250
Top Right - $25
Bottom Left - $25
Bottom Right - $45
Page 188
Top - $75
Middle - $100
Bottom - $40
Page 189
Top Left - $25
Top Right - $125
Bottom Left - $75
Bottom Right - $225
Page 190
Top Left - $125
Top Right - $55
Bottom Left - $200
Bottom Right - $45
Page 191
Top - $100
Middle - $75
Bottom - $225
Page 192
Top - $50
Middle - $800
Bottom - $125
Page 193
Top Left - $1,800
Top Right - $75
Bottom Left - $60
Bottom Right - $135
Page 194
Top Left - $225
Top Right - $250
Bottom Left - $125
Bottom Right - $75
Page 195
Top - $115
Middle - $75
Bottom - $60
Page 196

Top Left - $70
Top Right - $80
Bottom Left - $45
Bottom Right - $35
Page 197
Top Left - $35
Top Right - $125
Bottom Left - $150
Bottom Right - $140
Page 198
Top - $50
Middle - $70
Bottom - $75
Page 199
Top - $175
Middle - $75
Bottom - $275
Page 200
Top Left - $50
Top Right - $350
Bottom Left - $165
Bottom Right - $125
Page 201
Top - $140
Middle - $175
Bottom - $50
Page 202
Top - $175
Middle - $75
Bottom - $85
Page 203
Top - $50
Middle - $65
Bottom -$65
Page 204
Top Left - $65
Top Right - $60
Bottom Left - $130
Bottom Right - $75
Page 205
Top Left - $250
Top Right - $250
Bottom Left - $60
Bottom Right - $30
Page 206
Top Left - $230
Top Right - $45
Bottom Left - $350
Bottom Right - $160
Page 207
Top - $110
Middle - $110
Bottom - $65
Page 208
Top - $60
Middle - $100
Bottom - $275

Page 209
Top Left - $300
Top Right - $250
Bottom Left - $500
Bottom Right - $475
Page 210
Top Left - $300
Top Right - $325
Bottom Left - $175
Bottom Right - $35
Page 211
Top Left - $125 each
Top Right - $200
Bottom Left - $50
Bottom Right - $200
Page 212
Top Left - $225
Top Right - $300
Bottom Left - $35
Bottom Right - $75
Page 213
Top - $60
Middle - $45
Bottom - $50
Page 214
Top Left - $30
Top Right - $65
Bottom - $40
Page 215
Top - $55 each
Middle - $35
Bottom - $40
Page 216
Top - $75
Middle - $50
Bottom - $35
Page 217
Top Left - $75 each
Top Right - $150
Bottom Left - $60
Bottom Right - $75
Page 218
Top Left - $200 each
Top Right - (L) $250 (R) $175
Bottom Left - $75
Bottom Right - $50
Page 219
Top Left - $175
Top Right - $95
Bottom Left - $75
Bottom Right - (L) $125 (R) $95
Page 220
Top Left - $195
Top Right - $75
Bottom - $75